T0314021

Reliability Culture

Wiley Series in Quality & Reliability Engineering

Dr. Andre Kleyner
Series Editor

The Wiley Series in Quality & Reliability Engineering aims to provide a solid educational foundation for both practitioners and researchers in the Q&R field and to expand the reader's knowledge base to include the latest developments in this field. The series will provide a lasting and positive contribution to the teaching and practice of engineering. The series coverage will contain, but is not exclusive to,

- Statistical methods
- Physics of failure
- Reliability modeling
- Functional safety
- Six-sigma methods
- Lead-free electronics
- Warranty analysis/management
- Risk and safety analysis

Wiley Series in Quality & Reliability Engineering

Reliability Culture: How Leaders can Create Organizations that Create Reliable Products
by Adam P. Bahret
March 2021

Design for Maintainability
by Louis J. Gullo, Jack Dixon
February 2021

Design for Excellence in Electronics Manufacturing
Cheryl Tulkoff, Greg Caswell
January 2021

Lead-free Soldering Process Development and Reliability
by Jasbir Bath (Editor)
September 2020

Automotive System Safety: Critical Considerations for Engineering and Effective Management
Joseph D. Miller
February 2020

Prognostics and Health Management: A Practical Approach to Improving System Reliability Using Condition-Based Data
by Douglas Goodman, James P. Hofmeister, Ferenc Szidarovszky
April 2019

Improving Product Reliability and Software Quality: Strategies, Tools, Process and Implementation, 2nd Edition
Mark A. Levin, Ted T. Kalal, Jonathan Rodin
April 2019

Practical Applications of Bayesian Reliability
Yan Liu, Athula I. Abeyratne
April 2019

Dynamic System Reliability: Modeling and Analysis of Dynamic and Dependent Behaviors
Liudong Xing, Gregory Levitin, Chaonan Wang
March 2019

Reliability Engineering and Services
Tongdan Jin
March 2019

Design for Safety
by Louis J. Gullo, Jack Dixon
February 2018

Thermodynamic Degradation Science: Physics of Failure, Accelerated Testing, Fatigue and Reliability by Alec Feinberg
October 2016

Next Generation HALT and HASS: Robust Design of Electronics and Systems
by Kirk A. Gray, John J. Paschkewitz
May 2016

Reliability and Risk Models: Setting Reliability Requirements, 2nd Edition
by Michael Todinov
November 2015

Applied Reliability Engineering and Risk Analysis: Probabilistic Models and Statistical Inference
by Ilia B. Frenkel, Alex Karagrigoriou, Anatoly Lisnianski, Andre V. Kleyner
September 2013

Design for Reliability
by Dev G. Raheja (Editor), Louis J. Gullo (Editor)
July 2012

Effective FMEAs: Achieving Safe, Reliable, and Economical Products and Processes Using Failure Modes and Effects Analysis
by Carl Carlson
April 2012

Failure Analysis: A Practical Guide for Manufacturers of Electronic Components and Systems
by Marius Bazu, Titu Bajenescu
April 2011

Reliability Technology: Principles and Practice of Failure Prevention in Electronic Systems
by Norman Pascoe
April 2011

Improving Product Reliability: Strategies and Implementation
by Mark A. Levin, Ted T. Kalal
March 2003

Test Engineering: A Concise Guide to Cost-Effective Design, Development and Manufacture
by Patrick O'Connor
April 2001

Integrated Circuit Failure Analysis: A Guide to Preparation Techniques
by Friedrich Beck January 1998

Measurement and Calibration Requirements for Quality Assurance to ISO 9000
by Alan S. Morris
October 1997

Electronic Component Reliability: Fundamentals, Modelling, Evaluation, and Assurance
by Finn Jensen
1995

Reliability Culture

How Leaders Build Organizations that
Create Reliable Products

Adam P. Bahret

This edition first published 2021
© 2021 John Wiley & Sons Ltd

All rights reserved. No part of this publication may be reproduced, stored in a retrieval system, or transmitted, in any form or by any means, electronic, mechanical, photocopying, recording or otherwise, except as permitted by law. Advice on how to obtain permission to reuse material from this title is available at http://www.wiley.com/go/permissions.

The right of Adam P. Bahret to be identified as the author of this work has been asserted in accordance with law.

Registered Offices
John Wiley & Sons, Inc., 111 River Street, Hoboken, NJ 07030, USA
John Wiley & Sons Ltd, The Atrium, Southern Gate, Chichester, West Sussex, PO19 8SQ, UK

Editorial Office
The Atrium, Southern Gate, Chichester, West Sussex, PO19 8SQ, UK

For details of our global editorial offices, customer services, and more information about Wiley products visit us at www.wiley.com.

Wiley also publishes its books in a variety of electronic formats and by print-on-demand. Some content that appears in standard print versions of this book may not be available in other formats.

Limit of Liability/Disclaimer of Warranty
In view of ongoing research, equipment modifications, changes in governmental regulations, and the constant flow of information relating to the use of experimental reagents, equipment, and devices, the reader is urged to review and evaluate the information provided in the package insert or instructions for each chemical, piece of equipment, reagent, or device for, among other things, any changes in the instructions or indication of usage and for added warnings and precautions. While the publisher and authors have used their best efforts in preparing this work, they make no representations or warranties with respect to the accuracy or completeness of the contents of this work and specifically disclaim all warranties, including without limitation any implied warranties of merchantability or fitness for a particular purpose. No warranty may be created or extended by sales representatives, written sales materials or promotional statements for this work. The fact that an organization, website, or product is referred to in this work as a citation and/or potential source of further information does not mean that the publisher and authors endorse the information or services the organization, website, or product may provide or recommendations it may make. This work is sold with the understanding that the publisher is not engaged in rendering professional services. The advice and strategies contained herein may not be suitable for your situation. You should consult with a specialist where appropriate. Further, readers should be aware that websites listed in this work may have changed or disappeared between when this work was written and when it is read. Neither the publisher nor authors shall be liable for any loss of profit or any other commercial damages, including but not limited to special, incidental, consequential, or other damages.

Library of Congress Cataloging-in-Publication Data

Names: Bahret, Adam P., 1973– author.
Title: Reliability culture : how leaders can create organizations that create reliable products / Adam P. Bahret, Apex Ridge Reliability, Massachusetts, USA.
Description: Hoboken, NJ, USA : Wiley, 2021. | Series: Quality and reliability engineering series | Includes bibliographical references and index.
Identifiers: LCCN 2020032930 (print) | LCCN 2020032931 (ebook) | ISBN 9781119612438 (cloth) | ISBN 9781119612445 (adobe pdf) | ISBN 9781119612452 (epub)
Subjects: LCSH: Quality control. | Reliability (Engineering) | Corporate culture. | Leadership.
Classification: LCC TS156 B335 2021 (print) | LCC TS156 (ebook) | DDC 620/.00452–dc23
LC record available at https://lccn.loc.gov/2020032930
LC ebook record available at https://lccn.loc.gov/2020032931

Cover Design: Ana Faustino
Cover Image: drawn by Adam Bahret

Set in 9.5/12.5pt STIXTwoText by SPi Global, Pondicherry, India
Printed and bound by CPI Group (UK) Ltd, Croydon, CR0 4YY

10 9 8 7 6 5 4 3 2 1

Contents

Series Editor's Foreword by Dr. Andre Kleyner

The Wiley Series in Quality & Reliability Engineering was launched 25 years ago and has since grown into a valuable resource of theoretical and practical knowledge in the field of quality and reliability engineering, continuously evolving and expanding to include the latest developments in these disciplines.

With each year engineering systems are becoming more and more complex with added functions, capabilities, and longer expected service lives; however, the reliability requirements remain the same or even grow more stringent due to the rising expectations on the part of the product end user. With the new generation of transportation systems, such as autonomous vehicles, the expectations have grown even further. It will require the highest degree of reliability to convince people to entrust their lives into the hands of a driverless vehicle. Only with new visions, methods, and approaches to product development will this become a reality.

The book you are about to read is written by an expert in the field of product development and reliability and provides the methodology, guidance, and suggestions on how an organization should evolve to make a transition to the next level of maturity in regards to reliability-focused culture. It is my pleasure to introduce the author, Adam Bahret, a reliability consultant, who during his professional career wore a number of engineering and management hats, giving him the perfect opportunity to see the "big picture" of how product reliability is handled in various organizations. He has built upon this experience to produce a recipe of how to develop a reliability-focused corporate culture and emphasizes the important role management and leadership plays in this process. It is important that product reliability becomes the objective of the whole organization and not just an afterthought of a design process. Even though the ultimate goal of any organization is to deliver a fully functional, reliable product to the hands of the consumer, the intermediate goals and objectives may vary between the different parts of the organization, and it is the role of the leadership to align these objectives to achieve this ultimate goal.

However, despite its obvious importance, quality and reliability education is paradoxically lacking in today's engineering curriculum. Very few engineering schools offer degree programs or even a sufficient variety of courses in quality and reliability methods. The reason for this is hard to explain. Perhaps engineering schools prefer educating their students on how the systems work, but not how they fail? Therefore, the majority of reliability professionals receive their training from colleagues, professional seminars, and technical publications. This book is another step in closing this educational gap and providing additional learning opportunities for a wide range of readers with the emphasis on decision-makers.

We are confident that this book, as well as this entire series, will continue Wiley's tradition of excellence in technical publishing and provide a lasting and positive contribution to the teaching and practice of reliability and quality engineering.

Acknowledgements

I would like to thank my wife, Beth, and daughters, Katie and Natalie, for being so supportive during the creation of this book. I know many authors thank their family. I suspect many do it from guilt of having used so much family time to work on the book. But in my case, it was not just that. They really helped me when I was frustrated or needed feedback and a sounding board. It's love if someone listens to reliability engineering content and isn't in the field. So thank you for so much, love.

My sister, Abigail, was a great virtual coffee pot colleague. We live on separate coasts but would meet for coffee online and commiserate about writing. She is a script writer and I'm an engineer pretending to be a writer. She knew all the tips and tricks for getting through the hard times. You're the best, kid sister.

My good friend and writing coach Mark Levy. Mark, you are a great friend and I am absolutely sure no executive in the world will read this book unless you have put your magic into it. The best way to describe our edit is to simply say, "That is the first time this manuscript was in English, and not Engineer." You really saved this from being a book that couldn't communicate what it wanted to say.

I would like to also thank all of my colleagues and customers. I am extremely grateful to the executive leaders that allowed me to experiment and develop the techniques covered in this book with their teams and organizations. Your willingness to explore made this possible.

Introduction

When it comes to product development, most technology companies understand the importance of reliability. In particular, the engineering teams usually have everything they need to design a reliable product, including the right testing tools and analysis methods.

At times, though, there can be problems: a product doesn't ship on time or, if it ships on time, it fails in the field. Usually, and perhaps surprisingly, these problems aren't caused by engineering. The engineers have done what they're supposed to do, given the circumstances.

The problems come from higher up. They're generated by the organization's hierarchy. That is, the leaders caused the product breakdowns through their leadership decisions. Or, more precisely, they caused problems through *bad communication* of their decisions, and sometimes simply by making the wrong decisions.

For instance, when it comes to talking to their product team, a leader might give each role on the team a very different goal. They might tell the project manager that their goal is a launch at the end of Q1. . . and the R&D engineer's goal is a particular new killer feature. . . and reliability's goal is a reliability of 99.99%.

What happens then? Conflict! The project becomes a scramble.

Instead of the team collaborating on the overall product, they're competing with one another in narrow, limited ways. Each part of the team needs to achieve their particular goal at the expense of the other program goals. (After all, it's personal. Each member thinks, "I've got to win. My job is on the line!")

The result is a product launch that's ridiculously unbalanced. The product may hit the market on time, but the important new feature fails in the customer's hands.

What's more, the company has lost the opportunity to have a product that, originally, had promise, and the leaders have lost the opportunity to become the owners of a successful product program.

What I've done, then, is to try to right these wrongs.

I've written this book specifically for senior leaders. It's for those of you who want to achieve the goals of a highly reliable product, released on time, with the best new features. It'll show you how to build a culture that can generate impressive, product-based profits – while that culture is simultaneously centered on reliability. (If that sounds paradoxical or impossible, read on!)

When it comes to creating and releasing products that are innovative, popular, and financially successful, there's an odd belief out there. Leaders think that when it comes to weighing factors – such as cost point, time to market, and product features – there have to be harsh compromises. Fortunately, that's not true.

Often, the reliability advancements that will help you produce a home-run product are "free.'" In other words, if you more effectively connect the engineering tools being used in the program with your company's business goals, the reliability initiative will pay for itself. No additional costs or length. Hence, "free" in investment and profitable in return.

Before we go further, I'd like to tell you why this work is so important to me.

Consider the following scenario: imagine you're responsible for developing not just any product but one with life-and-death consequences. Let's say your product is a surgical device that cannot be allowed to fail. No way, no how. Failure means human death.

Yet what happens? The product development program shortcuts the reliability process, knowingly. On top of that, every announced schedule compression and budget cut strikes the reliability team first.

The consequences of such behavior are obvious, foolhardy, and painful. I mean, the FDA has walked in and shut down the operation before. History it seems is about to repeat itself. But why? Why are we headed into this horrible situation again?

In real life, I've experienced similar scenarios many, many times before, from nearly every side. As a design engineer. . . a reliability engineer. . . a reliability manager. . . and a leader who built entire reliability departments from scratch.

I've also seen it throughout the years as an independent reliability engineering consultant working on numerous projects in parallel in multiple industries.

We so often do things that just don't get good results. But we do them in that same way over and over again. The reason? We don't know why what we're doing isn't working, so we don't know how to do it differently. Change, however, is possible. We can understand why we're messing up, and we can learn to do things not just differently but correctly.

In this book I'll show you the "whys" behind common reliability mistakes, and I'll also show you a better way. Or even many better ways.

I created most of the tools and techniques you'll read about here, but they didn't come from inspiration. As my wife will tell you, I'm not some genius. Instead, I created them simply by having walked more miles in more types of shoes than most. I also keep my eyes open, tend to obsess, and am tenacious. These factors have yielded solutions that pack a formidable punch.

As a reliability consultant, I have the unique position of being a "trusted advisor," not only to an organization's management but to its executive leadership, too. Many of these leaders have been willing to give me latitude on the methods and strategies that have yielded great results. I'm appreciative for their willingness to go on these adventures with me.

What you'll find in this book are simply the methods I've seen work to ensure the product you develop is the product you planned.

"Reliability culture" is a study on how strong reliability and profitable business connect. More specifically, it will teach you how reliability plays a role in product success in the field.

The culture required to develop the Mars Rover's robotic assembly is very different than the culture needed to develop a toy robot for this year's holiday push.

When the Mars rover program began, the design team determined that each one of the vehicle's DC motors must have a reliability of 99.99999%. Why that extreme? Because if those motors made it to Mars and even one failed, the entire $2.1 billion mission would be a washout.

From a reliability standpoint, what that meant was that every single design decision the team made had to be done with the reliability of those motors in mind. Nothing could compromise those motors – not scheduling, not budget, not extra features.

That toy robot, on the other hand, better make it to the store shelves by third week of November or sales will be cut in half. If the reliability isn't perfect, so what? Most kids will have forgotten about the toy by Easter.

Until now, the reliability engineering discipline has been heavily focused on improving design processes. The problem is that design processes are only half the story.

The methods in this book come from having the ability to take a step back and connect the pieces. In reading how they work, you'll be able to sit down with your team and discuss the ones that will create the product you intended, are in line with your brand, and gain the company its greatest market share.

This is the beginning of an exploration into a new type of product development process. One that will allow your products to meet their full potential.

Let's look at the book's flow, so you'll know where we're headed.

Chapter 1, The Product Development Challenge, is an overview of common program difficulties. These include budget and schedule compressions that force leaders to cut necessary program steps. Such omissions leave management to make decisions blindly. The major factors that typically drive program decisions are outlined in detail. These product factors and how they interact with the program are critical for a successful project execution.

Chapter 2, Balancing Business Goals and Reliability, is about a major conflict: the fight between long-term and short-term business gains. The relationship between the modern business model and the reliability toolset is complex. Modern business methods want returns that happen fast, while reliability methods go for strong performance over the long term. See the conflict? These two don't match in strategy or execution. What can we do? Is there a way to bring the short-term and long-term thinking together, and make the result work for everyone? There is. You do it by cutting out some gut decisions and replacing them with decisions made quantitatively.

Chapter 3, Directed Product Development Culture, is about what drives organizational behavior. By exploring culture both inside and outside of an industry, we can dissect how it works and how it can be changed effectively. Just as important as change, is ensuring that the change takes root, so it can't be displaced by the "normalizing" forces of the daily operation.

Chapter 4, Awakening: The Stages to Mature Product Development, is about identifying where ownership and accountability lay for specific functions. The chapter discusses language and how teams communicate. By identifying the intent of language and opening the paths to direct communication tremendous jumps in effectiveness can occur immediately.

Chapter 5, Goals and Intentions, identifies the real reasons we do what we do. Without knowing why we are doing the program activities we are, we can't clearly connect value. Many activities have clear justification for the significant investment they require when we start a program, but the reasons get a big foggy as we progress through a program. Little can be achieved without goals. Even with defined goals, they can still fail to direct us if they are not well documented and in an accessible format.

Chapter 6, New Roles, outlines three necessary roles for a product development program. These roles, "reliability czar," "facilitators," and "change agents" are necessary if the correct accountabilities and paths of information flow are going to exist. These roles are not necessarily new hires for a department but an additional "hat" existing team members can wear at given times.

Chapter 7, Program Assessment, is about methods for evaluating your product development program's performance. By defining and measuring key success factors, we create a closed feedback loop to manage our program actions.

Chapter 8, Reliability Culture Tools, introduces the fundamental tools that ensure your product reliability culture is connected to the business's goals. Methods like "Bounding" and "Focus Rotation" are described in terms that allow the reader to implement them immediately.

Chapter 9, Guiding the Program in Motion, provides an overview of tools that can assist with keeping a well-directed program on track. "Guidance Bounding," "return on investment (ROI)," and "Program Risk Effects Analysis" enable the "closed Loop feedback" process necessary to apply regular and small course corrections so a smooth program gaining maximum return on investment occurs.

Chapter 10, Risk Analysis Guided Project Management, provides an overview of risk analysis tools and handling failure data. The different flavors of Failure Mode Effects Analysis (FMEAs) are reviewed with an emphasis on how they integrate into programs. Controlling failure data and root cause methods is fundamental to ensuring all the "free" knowledge floating around both in-house and the field are used to the greatest extent.

Chapter 11, The Reliability Program, discusses the strategies and implementation tactics you'll be using with your reliability tools. It'll guide you through the product and business factors that shape the reliability initiative. The major elements of a complete program plan are covered.

Chapter 12, Sustained Culture, is all about how to make your new reliability changes permanent. Change is good. Lasting change is how we win.

1

The Product Development Challenge

Rather than looking at concepts in the abstract, let's get down and dirty. I want to share a story with you about how product development goes wrong. In uncovering these traps, we'll then be better set up to talk about fixes.

What follows, then, is a kind of case study that highlights problems. And it's not based on a single company. Instead, the particulars are drawn from dozens of real life situations, which I've disguised.

Key Players

You and I work for Amalgamated Mechanical Incorporated. The company is hot about creating a new medical procedure robot. We're on the reliability team.

It's critical that our robot get out there quickly, because our #1 competitor is on the verge of introducing a similar product. To get a jump on them, our product needs to hit the market in 10 months' time.

According to the program plan, the Accelerated Life Testing (ALT) for the robot's arm should start next week. As far as we know, we expect to receive the arm samples in three months. If the ALT testing begins in three months, however, there's little chance we'll have accurate predictions on how and when the product may wear once it hits the market.

For all practical purposes, even if we provide that test-based prediction for the arm's wear-out failure rate a few months before release, it will be of little value. There's no alarm big enough to sound that would delay release, based on premature wear-out. Even if we discovered that the product wore out, not in the promised five years of normal use but in one solitary year, the leaders would still release on schedule. (After all, they'd reason, we have to beat the competition to market.)

It's been said before by our VP: "We'll release it now and get a fix out there quickly. We already have a punch list for version 2.0."

In many ways, the project has been designed to fail. For me, it feels like trying to stop a freight train that's built up a head of steam. Stepping in front of it creates a mess, and the train still pulls into the station on time.

Reliability Culture: How Leaders Build Organizations that Create Reliable Products, First Edition.
Adam P. Bahret
© 2021 John Wiley & Sons Ltd. Published 2021 by John Wiley & Sons Ltd.

Half the program's reliability testing was to provide input for program decisions:

- "How confident are we that the arm will reach its life-and-reliability goals?"
- "What's the robot's end-of-life failure mode?"
- "Should we create a preventative maintenance cycle or shorten the robot's promised life to customers?"

This is critical information in a product development program and we don't have it yet.

Why does it always go this way? It's actually made me think about changing disciplines to something other than "reliability engineering". Why have a career focus that doesn't improve products and is often just a check-the-box nicety?

The product did indeed release on time. The reliability growth (RG) testing showed low statistical confidence in the goal reliability. This is the most critical assembly and we don't believe it will work as it should, and we're going to release it anyway. This is crazy! The ALT testing was never finished, because there was a design change and we didn't receive new arms to test. So we don't know when it'll wear out. How scary is that?

These arms could start failing in large numbers in the customers' hands, because of a predictable wear-out failure mode. Statistically, the majority of the population will fail at this point, with a nice bell curve outlining the full population. More than half of the Failure Mode and Effects Analysis (FMEA) high-risk actions weren't addressed. Some of these actions had "user harm" in the severity ranking.

We released on time. About four months after product release, the field failure rate began to spike. Two specific failure types were dominant. The linear *X* axis bearing and a plunger that penetrates a consumable. Both see high cycles in use, and both were known to be high risk due to changes in the most recent design.

These spiking field failures were the main topic in every Friday steering meeting and hallway conversation. If someone had information, the CEO wanted to hear it. If there were no updates on the root causes and fixes, he yelled about wanting to know what everyone was doing all day.

For a reliability manager, the entire process was depressing. No real value was delivered from our team's work. As a matter of fact, we were usually seen as a nuisance – almost as if we were an outside regulatory organization, but without authority. Something akin to a kid on a Big Wheels pulling over highway motorists and issuing traffic tickets in crayon.

Now that there are high field-failure rates, people are murmuring, "Well, there's no single person to blame for all these failures... but... aren't you the reliability team? Why did you let this happen?"

As I said, I find myself thinking of abandoning the discipline I love, because it can feel pointless.

Taking a step back, I thought about the program's reliability experience from all the other roles involved. The project managers received bonuses for releasing the product on time. The R&D engineers were promoted and assigned to top-notch programs because of the features they developed. This was all celebrated at a fancy hotel with an offsite party and a band.

OK, that all happened at release. But what happened when the robotic arm assembly began to fail early in life? Surely, that was the moment of reckoning, wasn't it?

This is the team's experience when the failure rate spiked four months after release: they were called together as a "Tiger Team." This means they were borrowed from their new programs, because they were supposedly the only people who could save us: our "heroes!"

During this recovery phase, the Tiger Team got regular facetime with the CEO. Facetime with the CEO is a key element in someday climbing into upper management. Many of us would take a CEO face-to-face over a 20% raise. For the Tiger Team, this experience was pure gold.

Then, when the field issues were solved and the company was saved, there was a celebration with a festive banner and a big ice-cream cake.

As the legendary management leader Edwards Deming said: "One gets a good rating for fighting a fire. The result is visible; can be quantified. If you do it right the first time, you are invisible. You satisfied the requirements. That is your job. Mess it up, and correct it later, you become a hero" [1].

So, in summary, they were rewarded for casting reliability aside to enable meeting only one goal associated with their role: time to market, new features, or cost point. They were then rewarded again for fixing the field problems they themselves created.

Remarkably, the team was doubly incentivized to deliver a product that was unreliable. How could this be? Why did the program's executives engineer things this way? After all, it hurt them most.

But if I'm making it seem like everyone involved in this failed program was rewarded, I'm confusing the issue. People were indeed punished. Who? Those who really wanted to create a product that was reliable. What happens to those people? The next story shows that they have one of two paths.

Follow the Carrot or Get Out of the Race

The leadership of a large 90-year-old company asked me to evaluate their culture. In my report, I included a story. It had passion, reward, and, most importantly, punishment – all the elements of a Greek tragedy. So I wrote the story and got a response better than I'd ever hoped for. Here's the story:

I began my investigation with the question: "Why is reliability missing from most engineers' work, even though we promote it so assertively as a core value?"

Walking around the company's halls, I saw posters that underlined how seriously they took quality and reliability. These posters bore slogans like "Our customers count on us for reliable products" and "Our product reliability is YOUR legacy."

The hierarchy even jammed the word "reliability" as many times as they could into all their speeches. For instance, at the annual R&D off-site the CEO delivered an 11-minute talk, and used "reliability" eight times. That's almost one "reliability" per minute. I'd already been there long enough to understand the hypocrisy. That's why I was counting.

The company liked to hand out reliability awards. But these awards were largely empty. They rarely included bonus money or anything that resembled actual career growth.

It was easy to see management's true motivation. The late quality guru Philip Crosby said, "An organization is a reflection of its management team." There's no hiding what the boss truly values. Where this is most obvious is with things like sizable bonuses or meaningful promotions.

At this company, I saw engineers and developers rewarded when their product was on budget and on schedule. There's nothing wrong with handing out rewards for these accomplishments. Unfortunately, these were the only things for which the engineers and developers were rewarded. More notable accomplishments – like excelling at the full set of program objectives – were ignored.

To the team, upper management's reward-incentivized message was clear: "This is what we really value." The next level of management down had no choice, then, but to prioritize these same on-budget, on-schedule metrics above all others.

The written report I would later send to the organization's hierarchy had only two characters, Engineer #1 and Engineer #2. (Those weren't their real names. If they had been, it would have shown some amazing foresight by their parents.) These characters were a composite of the actual engineers on that team.

They differed from each other in a very important way. While they were both good engineers:

- Engineer #1 focused on budget and schedule, because she wanted the bonus and was ambitious enough to crave a promotion. She was attuned to what her management team valued, so she behaved accordingly.
- Engineer #2 followed her innate passion and decided to invest in reliability. Through careful analysis, she learned that her product had a reliability issue that could be resolved in four weeks. Of course, that would push back the product's release date by a month and cost $150 000 but she knew the fix would later save the organization $2 million.

From Engineer #2's perspective, when all was said and done her proposed fix would save $1.85 million, and she should have been rewarded for her thoughtful solution. But you already know where I'm going with this, and so did the leadership team reviewing her suggestion.

The problem with big companies, especially those with impersonal management teams sitting far away from engineering, is that the boss's directive of on budget and on schedule seems frighteningly rigid by the time it reaches the worker bees. The human resources department already has performance appraisal processes drawn up, based on budget and schedule. And because the directive came from so high up, there's not much scope for change at the lower levels.

So Engineer #1, who played ball with the leadership when the program was originally scoped, wins the bonus and a good performance appraisal. And they set an example for the rest of the workforce. They've become a symbol of good behavior.

What also happened, however, is that leadership told Engineer #2 (and everyone else watching) that this reliability stuff is unnecessary, even if it saves the company $1.85 million.

Engineer #2 will now disappear. If she's really good and prioritizes her professional happiness, she will simply leave and go to another company. If her value system doesn't align with the leadership's value system, and she has options elsewhere (and all the very best engineers do), she'll just disappear. That, or she'll simply turn into Engineer #1.

Anyway, that's what I put into my report (hey, they had paid me up front). These leaders, I wrote, had created a culture "allergic to reliability." If I were forced to be an employee at this company, I'd drop product reliability goals as a way of ensuring I received a paycheck.

The leadership had filled the company with people like Engineer #1, and all the opportunities to save $1.85 million were never taken.

Engineer #1 got promoted, but profits decreased and times grew tougher. And the organization is now conditioned to fixate on cost and schedule. It reinforces things that don't work even more. The culture gets worse. The talent leaves. Frustration reigns supreme.

It's Not That I'm Lazy, It's That I Just Don't Care

When the program initially started, the company's leadership talked about reliability as if it were life or death. They demanded "zero failures!" because "today, online reviews appear in seconds, and the market knows whether a product is good or bad instantly."

In the product-specification document, we set specific reliability goals. After hearing the leaders' demands, you'd think any reliability request would be answered with a blank check and permission to jump to the head of the resource allocation line.

Did everyone know they were going to abandon all of these high-minded reliability ideals? Boy, I hope not. That level of deception would be heartbreaking. I'm just going to assume they were as surprised as me when the reliability focus got derailed. But, really, let's be honest: I'm not that surprised.

Having said that, I still feel hopeful that the idea of reliability can take root within an organization. The reason I feel optimistic is that I believe everyone wants to build a product they can be proud of. No one wants to create a product that fails. If we take action that creates a product that's only moderately reliable, it must be due to pressure or objectives that aren't thought through. When teams have had an opportunity to create products with genuinely high reliability, while still achieving the other product goals, I'm certain you'll find nothing but enthusiasm.

That all workers want to create high-quality products was first discussed in Douglas McGregor's 1960 book *The Human Side of Enterprise* [2].

McGregor challenged the universally accepted principle that "workers don't want to work," which he called "Theory X." McGregor said that all workers want to do work that is meaningful and high quality. He called this "Theory Y."

We see symptoms that appear to support the lazy "Theory X" worker. But what commonly occurs are workers being pushed into work that they don't feel is meaningful. They're disconnected from the work's value, so of course there's little incentive to do a good job. This, combined with their not having input in how their role is executed, turns them into cogs in a wheel who only react to short-term positive or negative incentivization.

I believe reliability suffers from this same phenomena. When development teams are asked to do reliability activities in the product development process, they're disconnected from the actual value delivered. So they don't understand how reliability helps them achieve the results they want to achieve.

Sure, it's not as easy as showing them a diagram where A connects to B. It requires that they participate in the selection of the tool to be used. They have to believe that what they're doing is going to have an impact. If they understand how the tools work and select one they believe will have an impact, then they'll have created "a way forward." That's very motivating.

The three steps to making this connection between what they want to accomplish and the methods to get there are to educate, assign goals, and transfer control. The Design for

Reliability (DfR) requires that the full team understand reliability. Reliability can't be designed in if the designers don't understand the practices to measure and improve reliability in all phases of a product's life.

Previous to more expansive reliability education programs, the only reliability input that occurred in the program was from reliability engineers. This will never work if the objective is to have reliability designed in.

If you have kids then you know this dance. When you ask a five-year-old if they have to go to the bathroom, because you see them dancing a bit, they say, "No." This is even though they are about to lose control of their bladder. Sixty seconds later they will ask you to take them to the bathroom. OK, I get it: now it's your idea. Humans want to do things on their own terms from the moment they are born.

One of many issues with reliability engineers solely driving reliability is that the reliability engineer has to stand over the designer's shoulder and force themselves on the design process. Have you ever willingly accepted the input of someone standing over you? Me neither. The other option (usually what occurs) is to have the designers finish the design and then hand it to reliability to "make it reliable." There was an era that produced some evidence that this approach worked, and it's still referred to by old crusty reliability engineers today.

When computing electronics was new, in the 1950s and 1960s, there were a set of common best practices to making it reliable. If a reliability engineer took a new electrical design and then put these practices to the design, it would in fact be more reliable. But this is more akin to a copy editor cleaning up grammar and sentence structure for an author. That interaction isn't going to work with the development of new technology today. Not if you are going to be competitive.

Both of these strategies line up with the management style that was prevalent before McGregor shared his Theory Y. In this application, Theory X would be "designers don't care about reliability." That's not true, they were just being pushed hard on the other product and program factors. It was a management and cultural error that the designers' natural desire to make a high-quality product was not a part of their work. They also knew they had someone to throw it over the wall to make it reliable – so why worry about it? That was someone else's job.

By providing the designers with an understanding of the reliability toolkit, they can design in reliability themselves. It is not just the tools that increase reliability; it's also the tools that measure reliability. It's crazy to attempt reliability work without measuring product reliability during the process. How could you possibly know if you're not making a product reliable enough or, equally as bad, making it too reliable? Too reliable hurts, because the product will cost more than necessary and get to market slower.

The control part of this equation can also be tricky. By putting control of the product's reliability in the design team's hands, we're saying that the designers get to make decisions regarding release date, cost point, and product features. With that much control, if the team feels uncomfortable releasing a product on time, then the company would have to listen. Commonly, the company's concern is that putting this control in the engineer's hands will result in products that are never released. I've heard many managers say, "Engineers will try improving a design for ever. It's their baby and it can always be a better design."

To address this quandary, both the design team and the leaders need to understand the toolkit that reliability has to offer. By measuring performance and then comparing those measurements to the goals, decisions can be made quantitatively, not by feel. This is critical. The goals come from an agreed-upon balance of time to market, cost point, features, and reliability. It shouldn't be revisited every time resource requests are made. If that's the case, the goal never really was a goal.

When the leaders first set the goals, they likely put a lot of thought into those goals. They weren't random, based on a whim. You'll note I said the leaders have to understand the measurement tools as well. For two reasons. The first is, if they don't know a tool or measurement parameter exists, they can't request it. The second is, they have to trust the outputs from the tool. If the result is not understood, the result is questionable in their mind. Understanding where the value came from and being able to ask questions about how the method was executed are big factors in the credibility of the result in the recipient's mind.

So with the measurement being credible, and the goal being agreed upon by all parties, there is much less to argue about: "At Alpha, our goal is to demonstrate a five-year product life. The recently completed life test shows that up to 15% of a large population of our product may not achieve that life."

Unless we're going to argue about the original goal or the measured results, there isn't much else to do but address the design issues.

The title of this section "It's not that I'm lazy, it's that I just don't care" is a quote from a movie. Not just any movie but the cult classic for anyone who has ever worked in an office, 1999 movie *Office Space*.

The main character works as a software engineer at a company where he is just a cog in a mindless process. He becomes hypnotized and begins to just tell everyone how he honestly feels about everything. Coincidently, the next day the leadership has invited consultants to come in and find ways to improve the organization and process. Basically, see who they can fire and other ways to make cuts to improve the bottom line. Peter answers their questions with complete honesty. This is the full conversation between the hypnotized employee Peter and the consultant, Bob, interviewing him about his role. You can't tell me you haven't had a moment in your career that this moment does not capture perfectly. I'm grinning just knowing what I am about to write next, from memory.

PETER GIBBONS (SOFTWARE ENGINEER): The thing is, Bob, it's not that I'm lazy, it's that I just don't care.

BOB PORTER (CONSULTANT): Don't. . . don't care?

PETER GIBBONS: It's a problem of motivation, all right? Now if I work my ass off and Initech ships a few extra units, I don't see another dime; so where's the motivation? And here's something else, Bob: I have eight different bosses right now.

BOB SLYDELL: I beg your pardon?

PETER GIBBONS: Eight bosses.

BOB SLYDELL: Eight?

PETER GIBBONS: Eight, Bob. So that means that when I make a mistake I have eight different people coming by to tell me about it. That's my only real motivation: not to be hassled; that, and the fear of losing my job. But you know, Bob, that will only make someone work just hard enough not to get fired."

You're smiling right now, I know it.

Product-specification Profiles

When the program begins, we formulate the product objectives, which sound like this: "faster robot X axis motion of 10 mm $second^{-1}$, and a retail cost point of $23 500, and a first-year reliability of 99.9%."

Then, we feed these objectives into what ultimately becomes what's called a "product-specification profile." Until the product hits the market, this specification profile is supposed to be the project's touchstone. It drives all the decisions.

But how were the product objectives that went into the profile formulated in the first place?

Why, those came from the organization's business goals!

Really, everything we do at work is driven by the business goals. It might be one degree or five degrees of separation, but at the end of the day all the activity has to line up with the organization's business goals.

Your product's reliability, then, needs to be an adhered-to part of your business goals.

Do you see what I mean? Reliability is a part of your brand. What happens to the sales of all your product lines if your brand's reputation becomes tarnished?

It's not that every product needs to be 100% reliable, or that reliability is the sole factor in developing your product. You have to be clear about what you've decided, and stay true to it. After all, in many markets you're only as good as your last product.

As I've mentioned, a culture built on reliability is one where business and reliability connect. People often think of reliability as being part of the design process only. That's not true. It's only half of the story.

Executives at all levels understand how the reliability discipline affects market share. If the new coffee machine their company designed fails, they suffer personally. They're the ones held accountable.

Let's say this is the first coffee machine to market with Wi-Fi connectability, that it can email you when your cup of coffee is ready. This should mean it will grab significant market share and make the brand a household name.

But none of this happens, because it immediately has issues when customers try to get that desperately needed caffeine fix.

Uncaffeinated people who are denied coffee are quick to leave bad online reviews. Who cares about a coffee machine that can email you if the machine's water pump breaks when it is three days old? "Thanks for nothing."

So how is it even possible that a smart leader would blindly cut back on reliability initiatives when the schedule is tight or a new feature isn't ready for prime time? The reason is simple: when it comes to reliability, the direct effects on product or program performance investment aren't apparent – in the short term.

Businesses work on a quarterly system and individuals move to new programs and roles in short timeframes. Very few long-lead actions make it to the top of the daily action list. Most companies can't keep this high-level perspective when the day's urgent matters arise. Many that I have seen which hold this far-sighted perspective are privately held by a founder who was an engineer.

This lines up with McGregor's Theory Y, in that people want to do good work. But this type of situation occurs when "good work" is measured only in short, three-month increments.

An engineer who is also left holding the business accountability long term is the person who we should look to for best practices when long-term growth is the objective. This is one of the positions that seem to find a great balance for all the program and product factors. Let me repeat it to be clear.

An engineer with an accountability for long-term business success. I have proof of this. Some of the most successful companies in the world were founded by engineers who kept the business private. Some of the tools I propose aim to put the leaders in this same mindset.

Executives and owners of companies make decisions based on a significant amount of input from their council, their mid/high level management. If the executives and mid-level leadership's goals are not aligned then this council becomes muddied. This becomes compounded when the executives evaluate the mid-level leaders on the short-term delivery of time to market, cost point, and new features creation. The mid-level managers can't equally support reliability when times get tight and the results of an investment are far down the road.

So how do we fix this? We look at how executives make decisions. The first question is how do they get input/information for their decisions? Many executives work in a dashboard summary manner. The controls and readouts are very similar company to company, which is why it is easy for executives to move between organizations and industries. Reliability simply needs to find a way to format what is important so it can be included on this dashboard. And once created, this connection to reliability functions for executives will be similar to the other critical inputs executives receive day to day. Total Quality Management, The Toyota Way, and Lean Six Sigma have all had a significant impact on how we do business and product programs. They all have found ways to get on that dashboard.

Product Drivers

It was my first house, and it was a project like many first houses. Originally constructed in 1855, it had seen many caretakers. I was confident I would just be another name on the list and the house would be standing long after I was gone.

Being a first-time homeowner, I needed to begin that tried-and-true ritual of depositing my paychecks directly with Home Depot. Within a short time, I'd be meandering the aisles of the warehouse-size hardware store, with other homeowners, looking confused under the store's fluorescent lights, trying to figure out solutions to the problems that brought us here with little knowledge of how to solve them.

Before long, the cashiers knew me by name and welcomed me the same way the bar staff welcome the barfly Norm in the classic sitcom *Cheers*.

"Adam! How are the kids? Were you able to finish the ceiling?"

In a large hardware store like that, there are two types of people walking around: homeowners and contractors. From a product development perspective, these are two very different types of customers with very different needs.

As a "mortgage poor" person who was working on my own house, not because it was fun but because I was trying to save money, I wanted to buy tools that'd get the job done at the lowest possible price. Why lowest? Most of the tools I'd buy would be used a few times, and then would sit in my garage. My project was severely underfunded.

The other type of buyer, the contractor, was running a crew on a job site, where the highest operational expense was hourly wage. Having a tool fail would be devastating to the bottom line. It would leave workers sitting around. What that contractor needed to know was that the tool was going to work and not leave them paying idle workers.

So the power drill I'd buy and the power drill the contractor would buy had little in common other than both tools turned drill bits.

A product is defined by many factors. I find that the four primary ones are time to market, features, cost point, and reliability.

Imagine what I was thinking as I stood in front of the power drill display. There were at least a dozen models, and I had a clear ranking in mind as to what was most important to me: #1 was cost, #2 was cost, #3 was reliability.

That contractor standing next to me with the pro-wrestler-size biceps was evaluating these drills using totally different requirements: #1 was reliability, #2 was "This better not break!", #3 was "That quick-release chuck could save some time." Reliability was #1. Time-saving features were #2. Cost point was the last thing on his mind, because the expense of the tool failing was far more costly than the savings he'd enjoy in buying the lowest-priced drill.

Cost was my priority, because every dollar I spent on the drill hurt my bank account. Reliability was at the bottom of my priority list because if the drill broke I could work on something else and replace it the next day or even borrow a neighbor's. Reliability also fell low on my priority list because my duty cycle was a fraction of the contractor's. A design with low reliability used on a low duty cycle produces a similar failure rate as a design with high reliability and a high duty cycle.

I was doing many roles in my "job," and used the drill a total of one to two hours a day, every couple of days. The contractor may have a guy doing drywall construction all day – every day. In one month his drill might be used for 160 hours, while mine was used for 12. My project would be completed in three months, while the contractor works consecutive jobs throughout the year. That takes my yearly usage to 36 hours. The contractor's yearly usage could be over 2000 hours. That's why I can have a cheap drill and experience the same or even better reliability than the contractor purchasing the top-end model.

My whole point here is a drill is a drill, except that it's not. Each drill is designed differently, based on the factors that are most important to its particular market. Success isn't making the drill perfectly reliable. Success is making the right drill for the right customer, even if that means manufacturing a particular drill that has lower reliability.

The degree of a product's reliability should always be a conscious decision.

Bounding Factors

The design parameters I just talked about I call "Bounding factors." In a sense, they're the measurable factors that guide both your product and your program.

What are the Bounding factors for a product? A design feature or a cost point.

What are the Bounding factors for a program? Time to market or development cost.

The Bounding factors for a product and a program have to be related, because there are tradeoff decisions that affect both.

Here are some common Bounding factors we see in programs today:

- New technology/features
- Cost point
- Time to market
- Reliability
- Serviceability
- Manufacturability with a contract manufacturer
- Marketability.

The four that share the primary balance in most programs are those from the drill example: cost point, time to market, features, and reliability.

It's not possible to turn all of these up to a level of 10 (11, if you're a Spinal Tap fan), because they compete for resources. Extremely high reliability is not in line with quick time to market or ultralow product-development cost. New cutting-edge technology is not in line with low cost point or quick time to market, either. For those combinations, there's a give and take to get to a budget of time and money that works.

When this negotiating between schedule and investigating a reliability issue is done mid-program without program tools, it is often reduced to who makes the best argument at the time. I can think back to many programs where the decision was in favor of the person who was most in favor with the decision maker. "Who do I know best?" "Who do I trust?"

I don't believe the decision maker did this because of favorites. This would be insane, because if the program fails they fail. They do this because they don't have factual information at their fingertips. Without any quantitative information any individual is left with making a decision on counsel and trust.

Unless we can incorporate tools like the Bounding methodologies that will be shared in this book there is no way to expect a leader to make decisions on anything other than listening to trusted counsel.

The Bounding methodology derived its name from the base principle that each factor should "bound" resource and schedule changes to ensure no specific factor is compromised beyond the original product specifications' margins. What is being achieved is that factors are consistently steering the program toward the goals set in the product-specification document.

Reliability Discipline

Why is reliability done in product development programs? This is where we need to start. Without a "why" clearly defined we don't have a foundation to work on. The value in this early discussion is to understand this. Why is reliability test and analysis done? It is done to measure and improve reliability of the product. We usually stop there.

But why are we measuring and improving? This is the question that is left unanswered and we are without a way to make the critical decisions during a program. A great quote on "why" is by a sci-fi character named Merovingian in the movie *The Matrix*:

> "Causality. There is no escape from it, we are forever slaves to it. Our only hope, our only peace is to understand it, to understand the why. 'Why' is what separates us from them, you from me. 'Why' is the only real social power, without it you are powerless."

Let's change a few words in that quote to make it specific to our mission:

> "Causality. There is no escape from it, we are forever slaves to it. Our only hope, our only chance at program success is to understand it, to understand the why. 'Why' is what separates the good program choices from the bad program choices. 'Why' is the force that directs product programs, without it we are powerless. Making choices that are reactive, based on fear and blind trust are the 'why's' that make us powerless."

OK, that got intense. Let's go to the deepest "why" we can identify first.

The reliability discipline has progressed through several phases of maturity. It was originally approached as a method of identifying areas of risk and used "over design" as a mitigation. Look at any tool made from 1000 BCE to 1950 CE. I have many of my grandfather's tools, and they still work fine. More than fine, his tin snips could cut 0.5 in. (13 mm) steel cable. The wrench could be used as a hammer, and the drill could be pulled from the rubble of a house fire and used in the reconstruction. Take a look at the photos of his tools next to my modern-day equivalents (Figure 1.1).

But those product designs from many decades ago would be unlikely to survive in today's market. Looking at the modern-day equivalents next to them the differences are evident even to a non-tool enthusiast.

With rapidly advancing technologies affecting all of our lives, weight and cost quickly became critical to maintaining a competitive product in the market. The reliability engineering discipline took a more formal shape at this point due to a need to have a counterforce that ensured cost and weight didn't take the product design to a point of being, well, unreliable. The forces for product balance emerged: lightweight, low cost to manufacture, and at a low cost point are now balanced against reliability and developing new technology.

Through these needs, the methods of reliability analysis, test, and design techniques took shape. It was the military that led this initiative initially, simply because it is far worse to have a piece of equipment that will save your life not work than a vacuum or toaster quit at an inconvenient time.

Military customers still often measure reliability in terms of "risk of lost life." Even when a piece of equipment that is not intended directly for assault or defense (rifles, missiles, shielding, etc.) fails, it may increase the risk of loss of life. Something as simple as a surveillance camera on a tower in a hostile area can result in death if it fails and so needs to be replaced. A soldier doing maintenance on the top of a tower is a target. This places a whole new complexion on designers and engineers talking about a 30% failure rate of a camera in

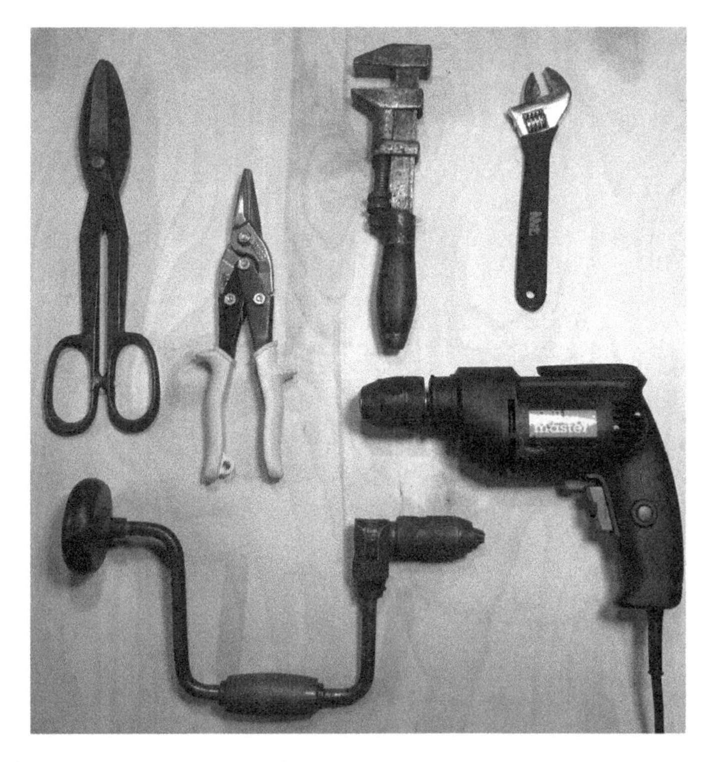

Figure 1.1 Grandfather's tools vs my tools.

a meeting back home, when that failure means not just that the camera stops working but that this, in turn, could lead to the death of a soldier.

I won't ever forget that sobering moment at that conference table when a DARPA general made that statement about a project I was working on. The clarity of what was at stake if I didn't succeed at the role I was brought there to do was chilling. When I had walked into the meeting I thought the consequence of a higher failure rate for the security camera was warranty expense and future lost sales. People's lives were at stake if a circuit board joint cracked.

The period for reliability from the 1940s to the 1970s was heavily analytical in nature. It encompassed tools like reliability predictions and specialized tests (Figure 1.2). The predictions were based on historical failure rates of individual components. The specialized testing aimed to predict the wear-out of specific failure modes or to identify the margin of failure of a primary stress. In the 1980s and early 1990s, more advanced techniques of testing like the Highly Accelerated Life Test (HALT) and Accelerated Life Test (ALT) became prevalent. These methods permitted more specific statements of reliability prediction or design improvement input to be made in very compressed timeframes early in the design process.

The 1990s to 2010 was very much characterized by the DfR initiative. This principle is that reliability is "designed in" not "tested in." A fundamental shift with DfR is that reliability practice is intertwined with the full team and design process from start to finish. This was a big shift from the mindset that mechanical engineers do mechanical design,

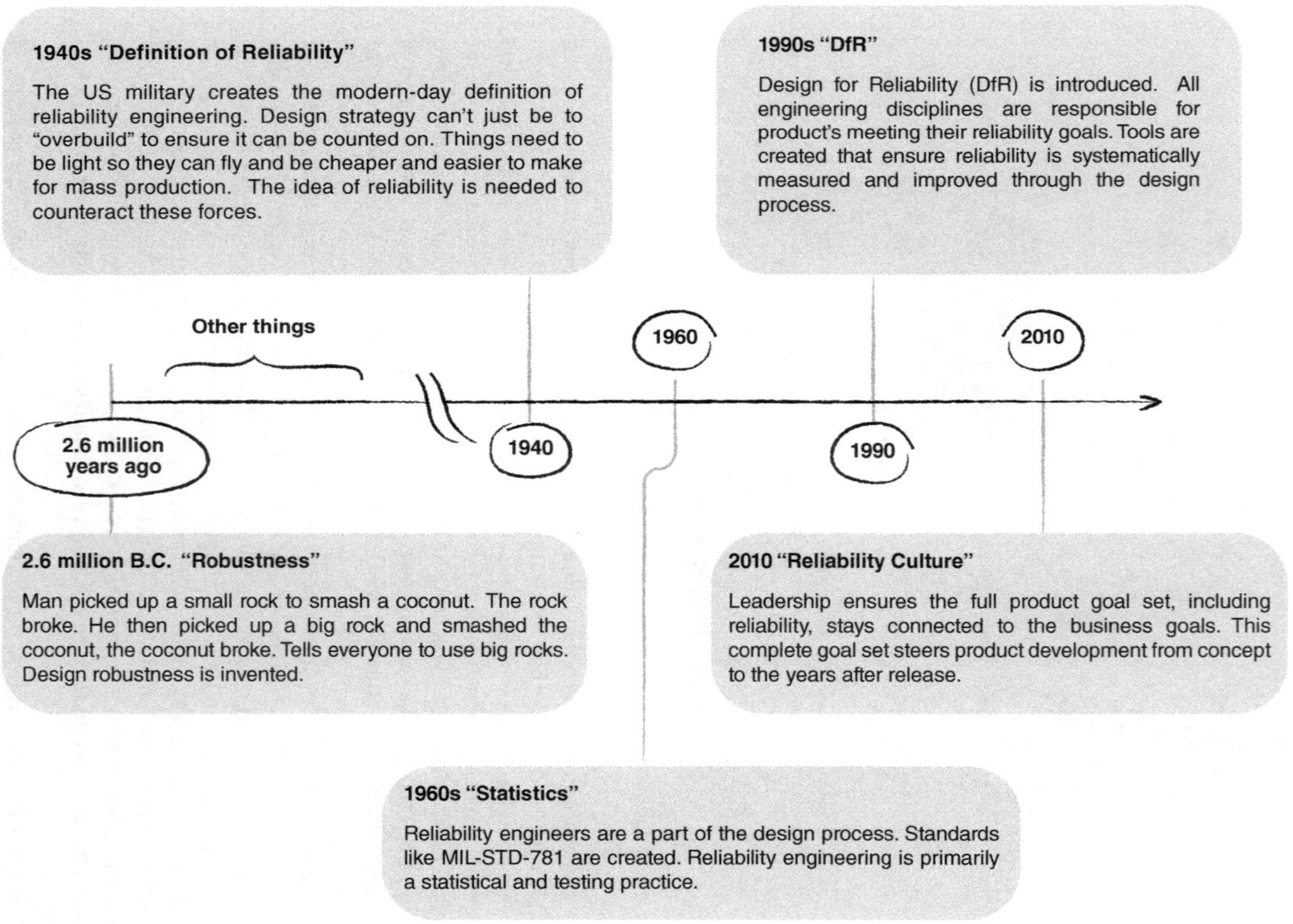

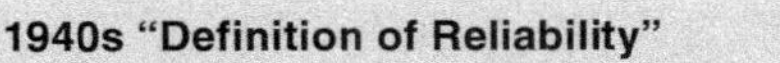

Figure 1.2 Reliability timeline.

electrical engineers do electrical design, and reliability engineers then make it reliable when they are done.

The next phase, today's phase of advancement for the reliability discipline, is reliability culture. This will be the connection of reliability tools techniques and philosophy to the highest levels of business and market objectives in conjunction with DfR. Companies that embrace this next phase of reliability evolution will quickly emerge as the leaders in their markets.

The bottom line in most of our businesses is dollars. The metric that will be used to measure an organization's level of reliability cultural maturity will be dollars. The return on investment of applied reliability tools will be measured in dollars. Companies that do not embrace a culture of product reliability will be ill equipped to compete with those that do. – just as it became impossible to compete without a Total Quality Management process two decades ago. We are on the cusp of placing reliability at the heart not just of the engineering process but of corporate culture. For this to happen, it will be necessary for business leaders to create the correct organizational dynamics and align reliability objectives with a business's financial goals.

Know your target. Make goals and make compromises. Don't commit to high reliability without selecting the sacrifices. Something has to give, be it schedule or new technology development or cost point or very high product development cost. There are no worse words than a leader saying "and it must be highly reliable" or "it will never fail" without discussing the cost of pursuing that reliability goal.

You have to know beforehand whether you are willing to trade reliability for growth of technology or time to market, and by how much. The Mars rover took many years and billions of dollars to create. From a technology standpoint the Mars rover is the equivalent of a high-school robotics science project. It has off the small digital cameras and a small DC motor like in radio controlled cars driving little wheels. There are servomotor-driven arms based on decades-old technology. I build stuff like this in my workshop with my kids.

But what was special about it is it could never fail. It truly was a "This design cannot fail or we are wasting billions of dollars." That was a quantifiable statement. The many years and billions of dollars spent to accomplish that perfect reliability was the cost. It would be a mistake to create your commercial or consumer product with that type of reliability goal. Nobody wants your perfectly reliable flip phone 10 years after the market has moved on to smartphones. So let's figure out what goals you should have for each program and how to correctly structure a program to accomplish them.

References

1. Edwards Deming, W. (2000). *Out of the Crisis*. Cambridge, MA: MIT Press.
2. McGregor, D. (1960). *The Human Side of Enterprise*. New York: McGraw-Hill.

2

Balancing Business Goals and Reliability

Return on Investment

It's difficult for reliability advocates to negotiate resources. Why?

Everyone knows that reliability affects sales, marketing, warranty expense, and future program resources. But those things happen downstream.

How can you compare them to arguments regarding return on investment for "time to market, "new features," and "cost point," which have such definitive returns sooner rather than later?

Arguments for investment in these types of factors are based on a short-term return that is immediately tangible: "For $25 000, we will reduce the cost to manufacture each unit by $4.25, a 12% cost reduction. This reduction initiative will be completed in 10 weeks, leading to a saving in manufacturing cost over two years of $2.3 million."

That's a clear request with an explicit return.

Compare that to the promise attached to investment in a reliability initiative: "We request $25 000 for testing, which in eight months will tell us we're 60% sure that we won't have a high failure rate."

"Huh? I get what for what? If I give you $25 000, I get to find out at the end of the project you're kinda sure things will be OK in the far future? Sign me up! No, just kidding. I'm giving that $25 000 to the cost reduction guy."

And that is what reliability faces on a daily basis. We offer "probability" in return for investment, not guaranteed cost savings, or a new feature that will increase sales, or getting the product to market six months faster.

Reliability engineers should start a support group that includes life insurance salesman and the guys with the signs on the street corner that say, "The end is near." We all have the same problem. It could be Ambiguity Anonymous.

Ambiguity makes it difficult to bring reliability into the conversation. The reliability work can potentially save you $15 million in lost warranty costs, missed market share, and brand tarnish. But I can't guarantee that or even give you a really accurate projection.

It's akin to selling seatbelts in 1940. "This new device is a restraint. It will cost you an extra $15 per seat, is uncomfortable to wear, and forces you to remain stationary inside the car.

Reliability Culture: How Leaders Build Organizations that Create Reliable Products, First Edition.
Adam P. Bahret
© 2021 John Wiley & Sons Ltd. Published 2021 by John Wiley & Sons Ltd.

But it may save you from injury or death at some later date." In 1940, a lot of people would have opted to use the extra cash to put toward a two-tone paint job, thinking, "I've never used a seat belt and I'm fine." (Well, the reason that's the case is because we don't often have the opportunity to suggest this option to people who've died in a crash already.)

But sometimes the person who has had the horrible experience of a loved one dying in a car crash will see the value in the $15 investment. Sound familiar? How many times do we see investment in reliability increase dramatically after a major product disaster?

Here, our mission is to get people to understand the value without experiencing the disaster first.

Change didn't happen until the department of transportation created a voice for all those dead people. They did it through marketing campaigns, based on the statistics of death and injuries in auto accidents. That approach worked, and today we know riding without wearing a seatbelt is crazy.

In any product and program, four factors are always present. They are:

- Time to market: for investments in dollars and working hours, we get a quicker time to market. That leads to a product with more time on the sales floor and more time in the market. When does the return occur? When the product is released.
- New technology and features: for our investment, we get features that will grab and hold new market share. When is the return? Months before release, we will know if they can be included in the new product.
- Cost point: we spend money on manufacturing techniques and design changes to reduce the cost in labor and materials. When is the return? We have a final number on per-unit cost by product release.
- Reliability: investment is in working hours allocated from other disciplines and initiatives. There is a significant cost in the program for test and analysis. When is the return? A year after release? Two years? That's far away and not even specific (Figure 2.1).

The people who lose out on this lack of investment more than anyone else are the individuals who set up the program and process. These are also the ones who pay the highest price. It's the executives that have to face the music when the products released are driving high warranty cost or customer dissatisfaction.

But somehow they're the ones that unknowingly structured a program that guarantees the customer doesn't get the product balance, which was so carefully planned when the project began. We have to conclude that they're not getting the information they need. No one consciously self-sabotages. This is the problem to be solved with our reliability culture initiative. "Why can't we understand the meaning of the information being presented at moments of critical decision?"

Program Accounting

One of the best ways to understand reliability's true cost is to study the practice of activity-based accounting. It's accounting that doesn't look at a product's materials and operational costs alone. Instead, it studies a product's complete cost.

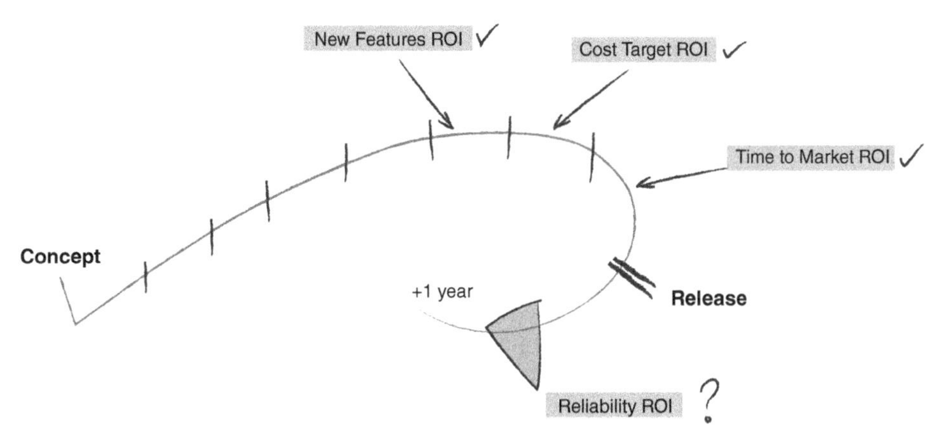

Figure 2.1 Reliability ROI.

I'll explain what I mean by citing one of my favorite things: donuts.

When you're 12 years old, a donut costs $1.75. When you're 45 years old, the real cost includes spending two hours on the treadmill to burn off the empty calories. Which also means I have to buy a gym membership.

The activity-based costing of the donut is now $1.75 + two hours of my time + a portion of my $125-a-month gym membership. Seen this way, donuts don't cost $1.75. They're expensive. I'll pass (or more likely I'll eat a few and pretend it doesn't matter).

Seriously, why do we humans do this? Let's look at a different cost path, bad health instead of the cost of exercising. Bad eating habits, like regularly bingeing on donuts, put me at a higher percentile risk for diabetes, heart attack, and knee and back problems from being overweight.

What is the difference in thinking between someone who knows all this "probabilistic" information and doesn't get the donut vs the one who does anyway?

This is exactly the same as the person who doesn't invest in reliability design robustness or reliability measurement because the outcomes are probabilistic.

Programs often budget based on what it costs to do something. "The cost of building four prototypes is $230 000." "The cost of developing this new technology is $2 million." Activity-based costing will measure the time a piece of equipment is not being used. That could be a significant add to both of those numbers. For a product development program this would be estimating the cost of having to do a redesign late in the program in comparison to doing that same redesign early. The initiative to include such assessments is often abandoned because of the difficulty of making it quantitative. "How much does it cost to do a redesign?" Not to mention this is usually occurring after the fact, and what's the point of counting how much money you have lost and can do nothing about? We also often don't do this evaluation early because "It's not going to happen to us."

Rule of 10s

The "rule of 10s" says simply: "To fix any design issue, you have to spend 10 times more than you would have if you had fixed it at the previous stage." In other words, the longer you wait, the more it's going to cost you (Figure 2.2). And the cost isn't going to be pennies or dollars. It's going to jump exponentially. If you wait long enough – through level after level after level of development – the fix will eventually cost you your shirt.

If it costs $100 to fix an issue in an early design, you can expect that fix to cost you $1000 at "design freeze," $10 000 at "first prototype," $100 000 at "product release," and $1 000 000 as a "field failure." That's a tremendous difference in cost to fix the same issue. These numbers aren't made up; they've been proven again and again. In fact, those of us who have been down this road are painfully aware that the rule of 10s is actually quite conservative.

How about a real-life example? If a team identifies a poor gear design early in the design phase, maybe a design review or sub-assembly test, it's a quick fix. In this case the engineer might, with some slight embarrassment, go back to his standard gear design practices handbook and make the correction. We can say that the fix took approximately a half hour of work, and it cost the program, say, $100.

Now, if this same gear issue drives a 10% field failure rate in customers' hands, we're in a situation that's painfully different. To get things back on track, you're going to have to call

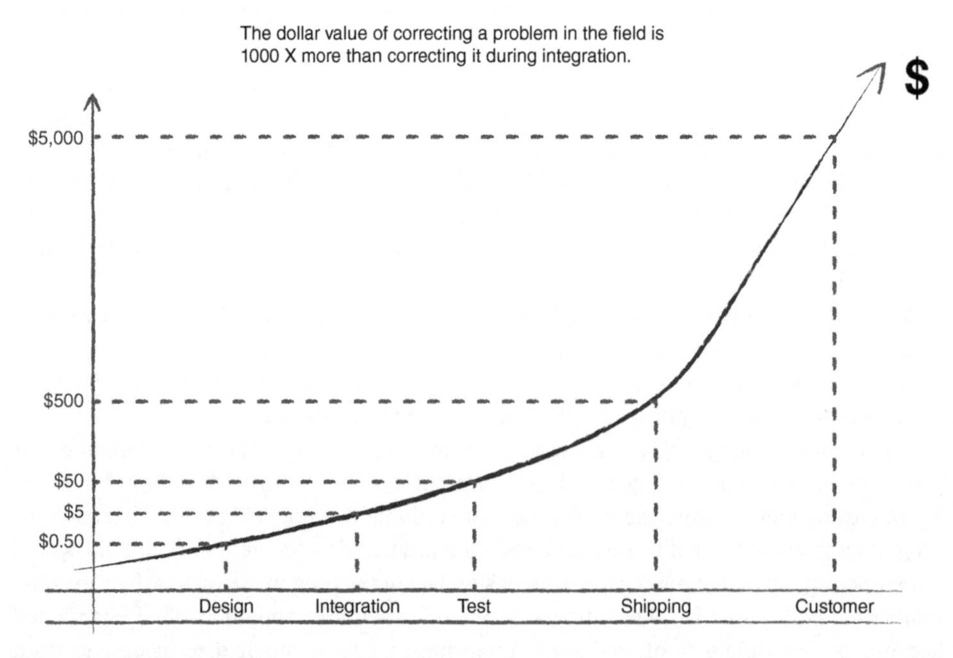

Figure 2.2 The rule of 10s.

in many people. The whole process will start with a root cause analysis, so you can get a handle on what's happening in the field. It'll involve the quality team, design team, and field representatives, at a minimum.

Once the issue is understood, we're still deep in the mud. That opportunity for the 30 minutes fix is far behind us. We're ready for the redesign process. But this version will likely affect surrounding components that were based on the original gear design.

OK, we figured out the issue, which is difficult in the field vs the in-house lab or design review. We're now ready to start the manufacturing of the new assembly. Once that's complete, we have to get it out into the field either through a full system replacement or on-site service.

OK, we did it and now we're staring at a number of disappointed customers. How do you think sales will go for the next quarter? "You're welcome, sales team!" Is it unreasonable to put a price of $1 million on this story? If you were to get away with all that for $1 million for a large-scale high-dollar product, I'd say you got off easy.

That's the true cost of not applying reliability tools early on. You needn't experience this kind of foul-up more than once to understand that the price tag on reliability tools and processes is ridiculously low in comparison to the events they mitigate.

An experienced program leader knows that an additional $15 000 of cost and four weeks of work to complete the Accelerated Life Test (ALT) was a smart investment, even if only 1 out of 10 ALT tests catch a major issue.

We clearly need to build a program that has the tools that'll stop us from hurting ourselves, again and again. (More on that later.)

Design for Reliability

So it's clear the savings in time and money are tremendous when reliability tools find issues early. Let's go a step further. What are the savings when we design in reliability from the start? This is the design for reliability (DfR) process.

Well, for one thing, we never have to find a faulty issue "in test," because it never existed. I'll reference one of the most significant studies done on the effect of DfR. Spoiler alert: DfR can easily cut program times in half.

The study was an academic study on liquid rocket engine development for space applications [1]. In this study they compared the complete set of steps of a traditional design approach, which focused on getting a workable design as quickly as possible, with a more robust approach, where the root causes of failure were designed out of the system before a prototype was even considered built.

Let's look at some of the major elements of the traditional design process in terms of cost over development time (Figure 2.3). The axes here are a vertical axis representing cost per unit time and a horizontal axis representing development time (schedule). The first element we will add is the actual design effort. This is what we most often associate with creating something: the blueprints, the diagrams, the dimensions. The next item to be outlined is engineering support. This is the people, the experts, the technicians, and everyone else who helps the design process.

Then the big one, eliminating failure modes. In this next area we will find the problems. These are problems that will affect either manufacturing deadlines or consumer reliability.

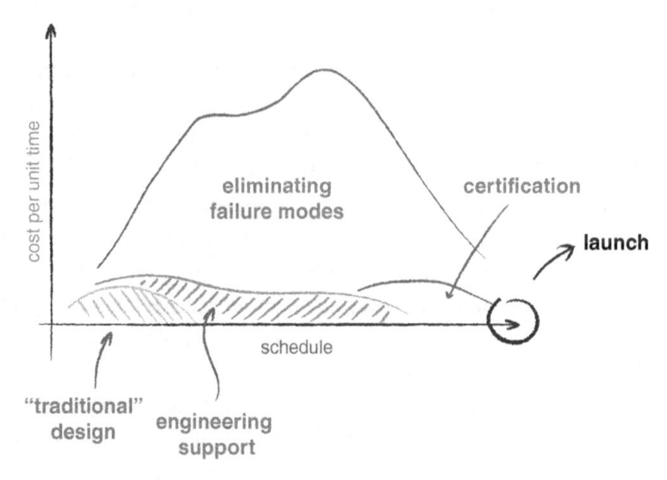

Figure 2.3 Traditional design process.

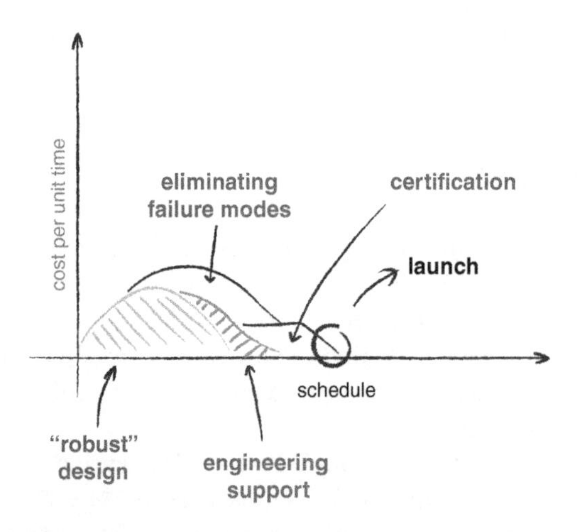

Figure 2.4 Improved design process.

These are often issues with prototypes under test – the issues we have to do something about before launch. The last element can generically be called "certification." What they are specifically will be industry specific. This also includes internal validation and verification procedures. After all that, we are ready to launch.

If the approach were changed, the design effort would have a stronger emphasis on creating a scientific (read "robust") understanding of what will fail before they start building prototypes. They put twice as much effort into just creating a working design. It is not only a working design but also a robust design.

They immediately found a massive reduction in the supporting efforts (Figure 2.4). The reason was a lack of failures and defects during testing. Why? They had worked most of

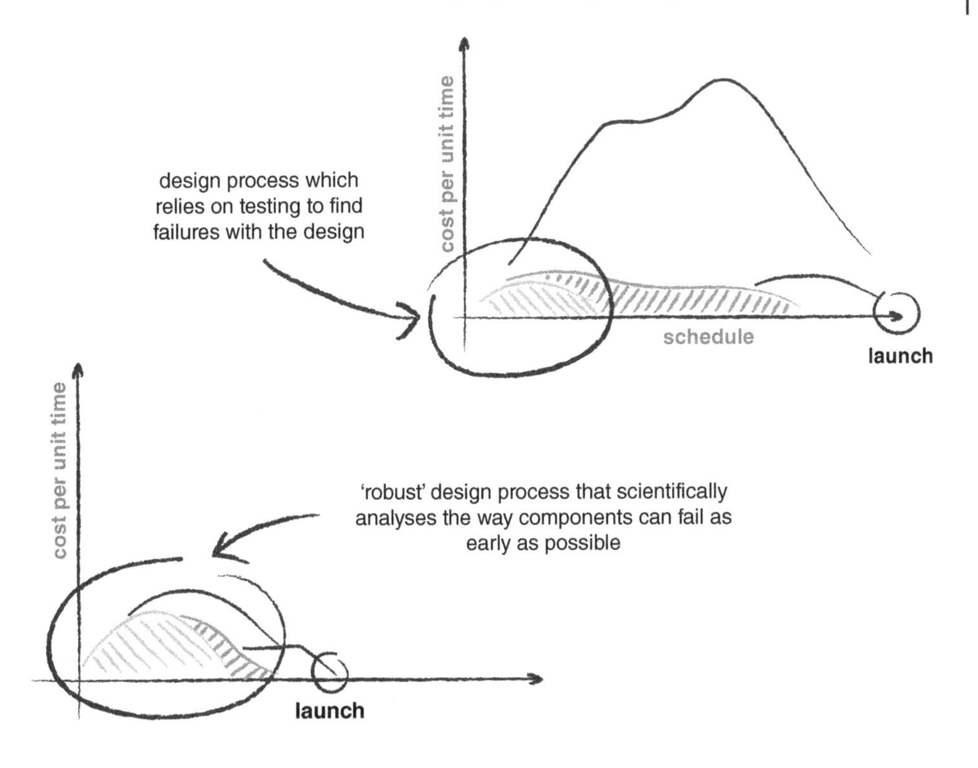

Figure 2.5 Comparing design processes.

them out before finalizing the design. This follows the rule of 10s. There was a much lower cost to removing the exact same issues by doing it earlier. Defects blow out production schedules more than anything else. So, by focusing on reliability, the launch date wasn't pushed out. It ended up being much earlier.

If we compare the two approaches (Figure 2.5), the robust process clearly has twice the raw design effort. But the total development effort has been reduced by 75%. Doubling the raw design effort means we spend much less time dealing with failure modes we weren't anticipating.

With a 50% reduction in development schedule, you have halved the overall development time. This is a reduction in time to market, which is what the people pushing back, the project managers, want more than anything else. They just haven't been measuring the true time to market when you consider the design recovery phase of the project.

Reliability Engineer's Responsibility to Connect to the Business Case

The challenges of keeping reliability at the resource and schedule negotiating table mid-program can be greatly helped by a solid effort to connect it to the business case. The reliability representative should spend time with sales, marketing, and leadership when the program starts to make these connections. All roles need to make an effort to keep this connection between reliability and the business case active.

What does the reliability leader do? A good reliability leader never loses sight of the primary program objective: "To gain and hold market share." They speak in these terms when outlining reliability strategy.

What does the organizational leader do? A great organizational leader always put this question to the reliability leader. "How does what you are proposing help the business?"

How hard is it to answer this question quantitatively? We tend to think it is too abstract a question, but it's really not. We just need to apply a few tools and we can easily do it. More importantly, we can consistently do it.

The relationship between these tools and the business case can be easily estimated. It just needs to be done early, at a more relaxed stage in the program: the beginning. These same people who are able to collaborate on this estimation early in the program may not be so willing when their day to day is covered in a red mist of panic. Simply, it's too difficult to create this framework mid-program when debates are hot.

It's easy for each of us to become hyperfocused on our individual role in a program. As I have pointed out this is often part of the dynamics created intentionally to incentivize us. Our leaders create a role to serve a specific purpose. "We need to design a mechanical assembly; get a mechanical engineer." "We need to keep the program on schedule and under budget; get a program manager." That purpose is why they were offered that nice salary and sweet signing bonus. "Do this function." But once they are on board there's all this talk about "the success of the team, the success of the company." But we all hear that voice in our head that reminds us why we were brought there and what we will be rewarded for. "I'm a PM. I'm going to crush that schedule. No one else has a chance!" Bonuses and promotions are tied to goals associated to a role. A product can fail and individuals can still be promoted or receive bonuses by accomplishing all aspects of the goals for their specific role. It's easy to find a scapegoat.

> "We really need to do something about Bob. He's killing us and has ruined the program."
>
> "Bob? Bob works in IT."
>
> "Yeah email was glitchy for, like, half the program. I can't work with this."
>
> "Congrats again on the new office. Nice window."

The executive leaders are responsible for the business. I rarely see this ownership transferred to individuals in the teams. This makes no sense, since the leaders are the ones directing the team. Because the goals aren't shared, the executives now have to manage at a high frequency. Would you ever walk your dog and throw treats to the left and right if you had a destination in mind? The leash would wear out with how much you would be pulling the dog back on the path. This is what they should be doing. If you want to get to a destination, put everyone's rewards together. "When we get to the park, I'll give you a treat." It's so they understand that the only reason you have the leash is so they don't get there ahead of you.

In the executive scenario it is exhausting to manage lower levels at a high frequency. There are too many layers and it's not the executive's job to do so. But when you see meetings with lower-level individuals "on the front line" giving updates to the executives you know we are in trouble. But we see this a lot.

To be honest, I really don't know how this weird and unproductive dynamic came about. People aren't crazy, and obviously you don't get to become an executive unless you are smart and productive. So there must be a reason. I think some of it is legacy. Industry changes fast, and factors like DfR aren't that old. Maybe there are some old paradigms we are still holding onto that are contributing to this.

Henry Ford once said, "Why is it that when I hire a pair of hands they come with a brain?" That is a really solid insight into the mindset of management when industrialization began. He also treated higher-level employees with a similar amount of disdain. Engineers had information withheld when they would design. He believed they were so beneath him that he had to unfold the project objectives to them in itty-bitty pieces so they could get through it. But this is the foundation of product development management we still stand on today.

We have obviously progressed greatly from this, but our deficiencies are still tremendous. This has been very strongly highlighted by the fact that The Toyota Way has been such a phenomenon [2]. Do you know what that book says? It's simple and can be summarized in a single sentence.

> "Every person in an organization has good ideas about how to do their job and improve the overall process."

This concept cannot be exercised in the same culture that has a "I'm getting what I want" base. In that scenario the voice of the assembly line worker who sees a better way to design a suspension control arm isn't going to make it to the design review.

So we need to change the "singular objective" that has been placed on so many roles. That's step one. I do acknowledge that we can't have a free love, no structure, hippy kind of thing either. We do have roles and they do have specialties aligned with an organizational purpose. We just need to mix in the right amount of high-level ownership of the end goals to each person, i.e. don't throw so many treats left and right as you walk the dog. There are enough squirrels and passing cars as distractions.

Let's analyze the singular objective. We need to start to mix it with the high-level goals if we want the balance of accountability that frees up leaders to lead and not waste time managing. I actually wrote "mix" but I think I'll change it to "blend." It sounds more measured and that is what we are going for.

It seems like I have been picking on everyone else, project managers, marketing guys, Bill the IT guy (he's so smug and I'm sure he is spying on me through my webcam). But let's look at how reliability can do better with this.

A reliability engineer can't consider the value of market growth due to a new feature introduction if they are concerned that any and all field issues are going to be indications of their ineffectiveness. How could you not be completely risk averse in that arena? Does R&D consider market share loss due to failure rate? They should, but not if a successful new technology introduction is how they are personally measured?

Our first guide for blending our goals is finding the business case.

Role of the Reliability Professional

The reliability professional's objective with proposing a request for reliability resources is to have the decision maker, "the buyer", negotiate only with themselves. It sounds strange to suggest your goal is to have someone negotiate with themselves. But in fact you have experienced this many times in your life. Have you ever requested something and then the opposing party simply gives you two choices? Who are you negotiating with now?

It's almost a Jedi mind trick. If you believe the information you have been given is valid "the truth" then you are really only left with making a decision. Why would you argue with the person who is delivering factual information to you?

I once saw a reliability manager get himself psyched up to go into a Friday steering meeting where he was going to demand that the reliability growth program be fully funded or he was going to put his badge on the table. I just rolled my eyes: "here, do you want to bring in a glass of wine to throw in the CEO's face (*Desperate Housewives* reality TV style) to make it a bit more dramatic?"

Why would the CEO agree to fund the program further? The reasons listed were just a bunch of empty doomsday threats. The CEO was also now in a position of having to stand his ground to maintain that he had authority. The debate was the CEO vs the manager.

This is how I do it.

> "We predict that we will increase our confidence in the reliability goal by 35% if we extend testing by four months. I also think we will improve the robustness of the design in the two main areas that have shown sensitivity. These areas have been improved to the point where no issues are occurring, but it would be great to further develop them. This reliability growth continuation will cost approximately $125 000, including working hours. If we conclude the program right now, we are under our goal statistical confidence by 23%. This doesn't mean the product won't be highly reliable – we see that it is performing well. It just means we are not sure because we haven't measured it for long enough. You will be able to get to market on your original production date if we conclude the reliability growth program now."

I never ask for anything. I'm just there to help make the best decision. If the CEO asks me what I would do, I will tell him, but only if he asks.

The CEO is simply left with an internal discussion/debate, with himself. Do I release the product on time and have an increased reliability risk because we have low confidence *or* do we extend release by four months and have the statistical confidence in the product that historically has served us well? He's debating with himself. The manager is only an advisor, a trusted and highly valuable advisor. Either way may be the right way to go. We, the advisors, don't know what all the other factors are in the big picture.

If the CEO believes everything that was stated as factual information then there is no further debate to have with the manager. The differentiator between this and the classical approach is that you, as the reliability leader, are not requesting anything. You are simply providing the buyer with the information they need to make a decision regarding what is best for the program. The difference is being able to provide information which aids that decision.

Quantitative risk assessment is finding a way to tie decisions regarding investment in reliability tools to the cost of unfavorable outcomes. The metric we commonly use is, of course, dollars. This is the only metric that has a universal adapter to all roles in the organization. The higher the role, the more dollars becomes the primary metric.

Historically, a reliability negotiation was between the reliability team and the program leader. The reliability team very emotionally campaigned for resources. These were requests for dollars and time. The request was based on nondescript doomsday scenario mitigations. "Remember on the last program when. . ." "If we don't do this then . . ." "I've been doing this for 30years and I don't understand why management doesn't make the connection between investing in these tools and having good products."

This is how children negotiate. They threaten and give hyperbolic doomsday scenarios. I know this because I have two very adorable terrorists who live in my house. The receiver of this type of campaign, me, has no other choice than to dig in and defend their position. When approached this way at home, I usually find myself starting off by saying "No!" to defend the attack, and then listen to the argument. At home or at work this really doesn't work well for anyone involved. In the end the program will make some small concessions to the reliability initiative. It won't be sufficient to get the desired results, and both parties will lose. Too little invested and nothing gained, the worst possible scenario.

A reliability advocate, at any level, will function like an advisor. Advisors don't argue. Advisors provide comprehensive information and opinions to the decision maker so they can make a good decision. They have to make a decision, and they are weighing the two choices with your counsel. This is one of the reasons that as a consultant I can find success much more easily than the internal reliability leaders. I'm viewed as an advisor.

As a consultant I engage with teams that have been having the same discussions for so long in the same combative format that it is the equivalent of trench warfare. I, on the other hand, can walk around the field and no one is shooting at me. I'm just there to assess what is going on, apply my experience to this one topic, and present the options and consequences each choice offers. I'm not there to judge what choice is made, and nor does it affect me.

If my role is simply to provide good counsel, a leader arguing with me makes no sense. When this happens, I am just sitting there wondering why they have asked (paid) for my counsel. As a reliability professional, become that person internally. If you become counsel to the organization, you can make change happen. How to do it? Stop making a case and start providing valuable information when decisions need to be made. The trick is creating information that is valuable.

The foundation to becoming that person is being able to put your opinions into quantitative facts. If you are expressing your opinion, the content is open for debate. If you are presenting the results of an analysis, the only real counterargument would be another analysis with differing results. If you are a trusted source of information then acquiring a second investigation is not a good use of time and resources. You are trusted and have facts.

The way to become this individual that transforms from opposing debater to advisor is to use tools that have been proven to correctly assess risk and dollars both pre- and mid-program. In Chapters 8 and 9, Where the Bounding and Program Risk Effects Analysis (PREA) tools are discussed, I review several risk assessment tools. In addition to being a great way to attach comparative risk to quantitative value, there is something else: "buy in." A cross-functional team has created this assessment, not a single person.

Once a large group of key individuals has completed an analysis, there is little a leader can argue with. Is a program leader going to argue the assessment that their entire team agrees upon is incorrect? If that occurs, you have bigger problems. It is critical that when this information is presented that it is concise. Do not make the mistake of presenting/offering the analysis itself. If these people are at the top of an organization, we can assume they are smart and assertive people. They will ask for detail, and how much, if it is needed for their decision. Oversharing the analysis details does two things that put you on a bad foot right from the start.

Oversharing how you got there creates confusion with regard to what information is critical for the decision. The second reason is that providing too much information does not bode well for credibility. Any sense of overload can trigger a feeling of "Am I getting smoke screened?" Feelings of confusion also create a sense of fear, which makes us less receptive to information.

When presented with a request for funding, an executive may ask, "These 33 samples and $125 000 for testing will get us to our 99.9% reliability goal?" A requester who is negotiating with them would say, "Yes, it will be a part of us getting there." There is no factual information in that statement. This person provided no information to assist with making a decision: they are simply attempting to make the decision for them.

An answer that "the advisor" would make is this:

> No those test will not get us to a 99.9% reliability. The 99.9% reliability will be designed into the product. ALT testing will assess when the dominant wear-out failure mode will occur. Simply, will the product last to the five-year goal we have set? If the test finds that the product cannot last five years, it will still be early enough to do something about it. Our options will be to redesign, add preventative maintenance cycles, or change the goal we are committing to.

The buyer can't argue with any of that. All you have done is provide them with factual information. What does ALT testing do? It measures something we need to know.

The buyer will then likely ask for specifics. "How many prototypes and how much is the ALT test?" The executive then has to weigh the value of the information that is provided by an ALT test and the cost of the units and money. The reliability professional is no longer a part of the decision, i.e. argument. The executive will negotiate with themselves either right there or over the next few days.

Summary

Early reliability integration into the product development process has been shown with a 100% success rate to be the most efficient and profitable configuration. But it is still not done that way in many programs. Why would companies not implement a proven method?

We have to look at the roles and how they relate to the program. The roles are where the actions are executed. Now that you have read this chapter, take a moment and think about your organization. Does the company hold reliability in a higher regard than individual roles do? How do the roles align with the company's objectives? Why would they differ?

This exercise will only have to be done for a few moments before a few "Ah-has" come to the surface. Share them candidly with the team. More than one leader has found that they have a team that is desperately wanting to "do it right" but haven't been able to because of role constraints. How wonderful would that be? Doing an exercise and having a few candid discussions and products begin to be developed better, quicker, and for less. I've seen that happen, more than once.

References

1. Rajagopal, K. (2005). An industry perspective on the role of nondeterministic technologies in mechanical design. In: *Engineering Design Reliability Handbook* (eds. E. Nikolaidis, D.M. Ghiocel and S. Singhal). New York: CRC Press.
2. Liker, J.K. (2004). *The Toyota Way: Fourteen Management Principles from the World's Greatest Manufacturer*. New York: McGraw-Hill.

3

Directed Product Development Culture

I have two goals for you. I want you to: (i) advance how you manage your reliability process, so it's efficient, and (ii) capture the highest possible return on investment (ROI) by creating a product that's balanced for your market.

With the way reliability is practiced today, however, reaching either goal is difficult. The question is "Why?" Why do well-intentioned reliability teams frequently fail in every type and size of industry?

The answer falls to a hidden force: your organization's culture.

Culture defines how an organization acts and works to maintain its status quo. Culture is like the organization's subconscious in how it directs our thinking and actions. Our culture implants within us the norms, and what's expected of us. It's not about overt rewards and punishments. Those are directives. Directives are inputs for conscious consideration.

Again, what we're talking about here is how the environment teaches us how to behave in a way that happens below the level of consciousness.

There's a theory that highlights how even slight changes in our surroundings give us unconscious clues about how to behave. That theory is called the Broken Window Theory. It goes like this: imagine you're in a crime-ridden area, and you see a series of buildings. They all look similar, except for one building. It's the only one sporting a broken window. The theory suggests that this single broken window will act as a trigger, signaling to criminals that the building itself is a target. It's more likely to have its other windows broken. It may even prompt more serious crimes.

That building's appearance is part of the block's culture.

Do you believe something as simple as a broken window can act as a catalyst for deviant behavior? I'm convinced, because I've seen the theory in action. More than once, I've been that deterred vandal.

It looks like this: it's nearly eight in the morning, and I've just eaten a piece of leftover pie. I need to rush out the door to my office, and I intend on just dropping the plate into the sink. I've told myself that, given my hurry, leaving the dish dirty makes sense.

When I walk up to the sink, however, and see it spic-and-span clean, without a single dirty dish, guess what I do? I rinse my plate and place it in the dishwasher. I even wipe down the sink to rid it of residual crumbs. The vandal has been deterred.

We're so easily influenced by our surroundings and their projected expectations.

Reliability Culture: How Leaders Build Organizations that Create Reliable Products, First Edition.
Adam P. Bahret
© 2021 John Wiley & Sons Ltd. Published 2021 by John Wiley & Sons Ltd.

By contrast, what do you think the result would have been if someone had demanded that I put my plate in the dishwasher? I might have said, "I'm in a rush. I'll get to it later" and then dropped the dish in the sink and run out the door.

For reliability to be an important part of the design process, the desire has to come from within each individual. Having reliability forced on people as a REQUIREMENT creates resistance. It's just human nature.

Now that we know what culture is, let's go deeper.

The Past, Present, and Future of Reliability Engineering

To begin our examination of organizational culture – so we can understand it enough to make it work for us and not against us – it's worth looking at our culture's very foundation. That is, rather than just thinking about the organization itself, let's look at where our organization is located geographically, and how that larger surrounding area and its history affects our day-to-day behavior.

In fact, let's begin by looking at a culture that's near-and-dear to the heart of the reliability engineer: Asian culture.

In your office I bet you have a copy of *The Toyota Way* [1]. If not, you could find a copy in a nearby office.

Within the broader Asian culture, the Japanese culture in particular has gone through several technology-and-design cultural periods that are interesting, reliability-wise. For thousands of years, the Japanese culture was known as an environment where objects and practices were perfected. One's personal identity and honor were associated with their trade. Fast forward to today.

Japanese products are some of the most reliable products available. In fact, that idea is such a truth that we commonly don't just purchase incredible Japanese inventions, we purchase our own inventions now perfected by the Japanese. Because, quite simply, theirs is a better version that won't let us down and will, at the same time, cost less.

There was, however, a brief period where Japanese products weren't highly regarded. From the mid-1940s until the early 1970s, products coming out of Japan were commonly thought to be cheap and disposable. You know, junk. In the mid-1970s, that perception began to change.

Influences

This transformation is definitely worth analyzing, because the recovery was so quick. What were the influences? What was the culture?

The first question we should ask is this: "Why does the Japanese culture, as well as other Asian cultures, have the core ability to perfect skills and maintain high quality?" Or specifically: "Why do Asian cultures perfect designs for cost, quality, and reliability?"

To answer this question, I'll pose a theory first read in Malcom Gladwell's thought-provoking essay "Rice Paddies and Math Tests" [2]. Says Gladwell: "If you look at Eastern cultures, a staple crop is rice. If you look at Western cultures, a staple crop is wheat. This difference in staple crops may explain why Easterners take time for perfection while Westerners find it easier to create ideas, designs, and technology that are brand new."

Gladwell made a correlation between the behaviors needed for rice farming with the habits for good math grades. We're all familiar with the notion that Asian students are better at math than Western students, right? It's a stereotype. But the data does in fact show that, in general, students raised in Asian households have higher math grades than their Western counterparts.

How might we explain the correlation? Both rice farming and math mastery require diligence and a "never stop trying" mindset. Both talents demand similar skills.

The following experiment was carried out to prove that, in mathematics, the quality that directly correlates to success is tenacity.

A math test was given to a group of students from mixed backgrounds. The test had several questions of varying difficulty. The time it took each student varied, based on their abilities. If you plotted the time it took the group as a whole, you'd get a standard deviation. So now we had time to completion of the test and grade for each student.

At a separate time the same group was given a new test. It had a single problem. But this single problem test was a different kettle of fish. Solving it wasn't only hard; it was impossible. The problem had no answer, only the test-takers weren't told that. What happened?

Secretly, the observers were measuring how long each student would work before they quit. In other words, they weren't really measuring people's ability to solve the problem. They were measuring their stick-to-it-iveness. The conclusion: there was a direct correlation between tenacity and actual math ability. Those who worked at the impossible problem longer had in fact the higher math grades on the real test.

We could conclude that, tenacity equals good at math. Rice farmer equals tenacity. We can then conclude that good rice farmers are going to be good at math.

Many of our cultural behaviors are rooted in practices handed down from our ancestors. These behaviors were passed along unintentionally, yet they profoundly shape how we operate in the here and now.

The Invention of "Inventing"

As a reliability advisor, you can imagine the (halogen) light bulb that lit in my head as I read Gladwell's idea. It explains why you and I try to make better products by reading *The Toyota Way*.

Since the dawn of time, I suspect, the first technology developed in every society was to handle food production. As hunter gatherers, tracking, killing, collecting, and later cooking food over a fire were the activities that preoccupied us most.

Hands down, the greatest change in how we lived was the technology that allowed us to stop roaming and live off planted crops. It's no understatement to say that this was the single greatest event in the history of inventing, engineering, or science. Ever. It allowed us to invent at a pedal-to-the-metal pace that was insane.

Planting and harvesting crops was akin to inventing "inventing." This towering event meant two things: a tribe could stay in one place longer, which made it possible to create a simple infrastructure. Also, a single individual could produce enough food to feed more people than just their own family. That, my friend, was the big accelerant. Once that could be done, inventors could invent instead of gathering food. Their tools and work areas could be more elaborate.

This is "time and resources"; the same ingredients we need to create today. Here, for the first time, they were part of the equation. The triggered creation-explosion has sped up at a dizzying pace. It's faster now than ever before.

When my dad was alive – from 1927 to 1999 – all the technology stayed relatively consistent. It was cars, airplanes, typewriters, radio, and television. From the time he was a boy until he was in his seventies, he could count on being surrounded by the same kinds of technology he always knew. Sure, over the course of his life the machines he'd come to know so well advanced in complexity and quality, but they were essentially the same throughout the decades.

I have two daughters, born 2005 and 2008, and their relationship to technology is far different than my dad's. They know only one thing: change. Whatever technology they know today will surely be different tomorrow. The progression moves so fast that their generation can't be defined by any specific technological lifestyle.

When my 14-year-old daughter arrived on planet earth (2005), we still made phone calls using landlines. Today, not only are landlines ridiculous, but so are phone calls. If my daughter gets a call, her first thought is, "Who died?" If no one died then her next response is, "Why didn't you just text me?" When she was younger, my daughter burned her own CDs. When she wants to hear music now, she yells out the name of a song into an apparently empty room and the song plays.

Technologically, what will her life be like in a decade? Different. That's the only prediction you could make accurately.

And it's all because human beings planted crops.

Like most major inventions and discoveries, the planting of crops happened by luck and creativity. The luck part was in finding plants that could serve as crops. Believe it or not, out of the millions of types of plants on earth only 12 can be made into crops efficiently. Finding those allowed us to transition from wandering hunters to stationary farmers.

Actually, that was only the first element that allowed us to change. The next element was our understanding of how to handle crops through all their growth stages. This required tools for preparing the earth, planting, weeding, and harvesting.

Once it became possible for an individual to feed more than their own family, others could begin to specialize in skills that could be used to trade for food.

That's it, the prehistoric switch flipped. It was no longer necessary for every person to spend all day acquiring food. It soon became possible to be a "specialist." Specializing in a particular skill meant that technology through invention and practice could advance at a rate never before experienced. It was the dawn of the skilled professional.

The transition to farming led to the invention of rope, plaster, ceramics, and metal alloys made from raw materials. The combination of suddenly needing more technology and having the time to create it was the cause of the first technology explosion.

If a community lived in a geographical location suitable for rice, the behaviors and tools developed would be shaped by the needs of rice farming.

Quality and Inventing Are Behaviors

If you've ever seen a rice paddy up close it's an amazing system. The complexity and finesse needed to produce a strong yield are impressive. Any neglect of the crop during its growth cycle, and you risk losing the entire paddy.

A rice farmer is tasked with creating a paddy that has a foundation of clay and soils. The paddy is dimensioned in a very particular way, and the field has water gates to ensure the correct water levels are held. If the water is too low, you can lose the paddy to weeds. If it's too high, the rice will drown. The rice farmer must monitor all these parameters and many others and make adjustments daily. If a rice paddy is neglected for days, the crop's value could quite possibly be lost.

An old Chinese proverb reflects the virtue needed to tend a rice field: "No one who can rise before dawn three hundred sixty days a year fails to make his family rich."

Clearly, in this culture the key to success was the virtue of daily diligence.

When it comes to rice, the behavior rewarded could rightfully be termed "quality control." The reward is life. The punishment is starvation for you and your family. The stakes are high and the customer, rice, is harsh and accepts no excuses. If you look at the arrangement between the rice and the farmer, the relationship is clear: "Tend to me day in and day out, make adjustments, and I'll feed you and your family."

You can quickly see how striving for perfection through careful monitoring, analysis, and correction would become a cultural cornerstone. This is exactly the mindset that makes for a good quality or reliability engineer.

Now, let's look at the staple crop for most Western societies: wheat. What does wheat farming look like?

To get the crop planted, you need a burst of effort. The faster you plant, the more acreage you cover. As the crop grows, there's a lengthy period where little can be done. That is the "wait and hope all goes well" phase. At harvest, you'll need a second energy burst. The faster you harvest, the greater the yield.

What types of behavior, then, does wheat farming reinforce?

If my challenge is to plant as much acreage as possible, and wait out lots of downtime, then harvest quickly, I'd say an "inventor" mindset fits. That's because I need to find creative ways to plant more. As the crop grows, I have time for new ideas. If I can invent ways to better plant and harvest, my family prospers.

Sounds a lot like the type of fast innovation characteristic found in many Western cultures, doesn't it?

Back to the Japanese culture: so why was there a period where Japan was known for producing cheap junk and then this amazing leap to becoming a leader in the methods for developing highly reliable products?

As Always, WWII Changed Everything

WWII, obviously, played a major role in that transition. If you looked at the world in 1945, when WWII ended, you'd see that most Japanese and European manufacturing facilities had been sought out as targets for bombing raids. These manufacturing sites built war machines, and eliminating them was the fastest way to disable an opponent.

What did America's manufacturing capabilities look like in 1945? They had gone through an unprecedented expansion to mega facilities that the world had never seen before. The US government poured billions of dollars into manufacturing. The automakers were all converted into tank and airplane producers, and expanded their size by four and five times. During the conflict, our enemies couldn't reach the American mainland, at least not in great numbers, so not a single US manufacturing plant was destroyed. After the war, the US had tremendous manufacturing capability and a vastly reduced need for war machines. That set up the beginning of US industrial domination. Fully capable, very little competition, and a huge new customer base.

For an inventor culture, this was the ultimate invention playground. Each postwar year, we went crazy turning out wild new technologies and products. The faster you got it out there, the better the sales. All that mattered was speed and innovation.

Following their defeat in WWII, Japan made whatever they could with whatever materials they could find. If you popped open a Japanese party noise maker, on the inside of the thin sheet metal you'd find soda can logos. Everything was made from minimum resources, and the item was only as robust as it had to be. If it wasn't supposed to be dropped then you'd better not drop it.

From our perspective, these products were cheap and untrustworthy. We like to drop things, pick them up, and keep going.

Meanwhile, our design strategy was to invent, advance, and overdesign for robustness. In the US that same toy would be made from thick stamped-steel, producing an item that was capable of being handed down from generation to generation. You can still find them in antique stores today. Good as new, minus a few dents and paint chips.

The US was in a golden era of new families (and customers), had a strong economy, and had the capacity to invent anything we wished. There was little global competition, because everyone else was picking up the pieces. We made cars sporting big fins, and why not? Every Chevy, Chrysler, and Ford had enough chrome in the bodywork to signal a passing spacecraft. We came up with all kinds of crazy ideas for products and put them on the market, pronto. It was about speed and invention. Be the first to bring this next new amazing thing to market. That was how you made a sale: invent!

The consumer was fine with the other side of this fast-and-loose product development. No one complained that it was common to spend Saturday mornings tuning or repairing the family car. Most of the car was so overbuilt that it would last for generations, yet no one had worked out the design's bugs yet. Vacuum cleaners went to the repair shop once a year. That's just the way it was. You had the latest greatest product, so who were you to complain in this grand era of "living in the future" because something broke?

The Postwar Influence Diminishes

But then something happened. W. Edwards Deming began to share a bunch of interesting ideas. Deming was an American who invented something and it wasn't a product. It was the quality process in manufacturing. American industry, however, wasn't interested. It viewed quality as a needless process that would just slow product development cycles. No customer cared about perfection. They wanted new.

When America wouldn't take him seriously, Deming went abroad with his ideas. In what would become a regular pattern, the Japanese grabbed onto a great American idea, perfected it, and brought it to the international market to change how things are done. But quality wasn't a one-off type of thing. It also fit some of Japan's most fundamental cultural philosophies. The "rice farmer way." Japan embraced these principles, and soon enough Western manufacturers were put on notice. In many ways, we still haven't recovered.

In the 1960s and 1970s, these little cars began showing up on US shores. Many Americans made fun of them. The US car companies dismissed them as niche novelties coveted by weirdos. The path to success for these little cars could have been much slower than it was. But an event that couldn't have been predicted accelerated their acceptance curve.

That event: the 1979 oil crisis.

Suddenly, cars with great gas mileage became more important than the American desire to drive a vehicle with a backseat the size of a sofa. For the little foreign cars, sales climbed. Drivers were buying the cars for the fuel savings, but they got an unexpected bonus: most of the cars kept going without extensive maintenance.

The Emergence of Japan

Fast-forward to today, and every American manufacturer reads *The Toyota Way*, which could have as easily been titled *The Rice Farmer's Way*.

Now on the other side of what at this point must just seem like a love letter to Japanese quality and reliability: through post-WWII to the millennium, the Japanese definitely aren't known for being leaders in invention. Many of the products we commonly purchase with Japanese brands on them were invented elsewhere. So there are some definite lessons they could take from the wheat farmers, and they have.

Japan is investing heavily in helping its designers develop an "invention" mindset. A few mindset hurdles they have had to address are that historically it is frowned upon to bring attention to yourself as being better than your peers. New ideas should be credited to the team. This type of environment is not conducive to brainstorming. The intense focus on perfection leads an individual's work into optimizing the same task, not evaluating if it can be done a better way.

The Japanese have aggressively added "invention" to the skills they develop in students and industry experts. Continuing to use the automotive industry as an indicator of product culture, we can see the struggle they have had over the past decades to shake the stigma of producing bland nondescript cars. It has been a long journey for them to capture the passion of automotive buyers who wanted more than a trouble-free way to get from A to B.

One of the first attempts they made to break free from this was the delivery of the 1969 Datsun 240Z to the shores of North America. I own one of these first cars that showed a determination to break this stereotype, like the dorky math kid who comes into school one day wearing a black leather jacket. Yes, he still has his slide rule on his hip and his pants are too short but he's trying as hard as he can to be cool.

A Datsun 240Z was indeed a sexy fun car. It was a big surprise to everyone.

But it was affordable to the everyday man, fuel efficient, and reliable. It did in fact have more horsepower than the Porsche 911 and handled better than a Jaguar E-Type, and at half the cost. On paper all the numbers were there, in true Japanese style. But it was still a little awkward when it was next to the legends, like that kid with high water pants and a leather jacket trying to smoke outside with the cool kids.

It didn't put Porsche or Jaguar out of business, or even take any of their sales, but it did show a self-awareness and drive to improve where they needed to: passion and innovation. It also brought great sports car performance to people who otherwise would never be able to afford it. This was their innovation.

I also own a Porsche 911 from the same era. It is a little too expensive for what it is, but there is something almost magical about it. The performance numbers aren't off the charts, but I can't imagine getting the same feeling in any other machine.

It is a car that melds with you and as your skillset increases it unlocks the next levels of the car's abilities. It almost grows with you like a coach. The Datsun is so clearly a copy of this intent. You can tell it is designed by overly analytical personalities who wanted to copy the driver passion aspects of European sports cars but also make it economical, efficient, and perfectly reliable. It's someone doing a cover of a song by your favorite artist. Enjoyable, lower ticket price, but not exactly the same. But you got 90% of the experience for half the ticket price. I drive these cars back to back just to feel exactly what occurred at that exact moment in time, like a gear head archaeologist.

Reliability Is No Longer a Luxury

Some moments in consulting will never be forgotten. When I was working with a product development company that specialized in consumer indoor electronics, there was one of these moments. It was when I heard how they do closed-loop field failure rate reports.

Online reviews make it to the engineer's desk in 24 hours. After understanding that I audibly said, "There are no more free passes."

Of course, there was a time when you had a free pass and could fix products in the field. Certainly, when Henry Ford was alive you could.

Ford did something amazing when he created the flathead V8 in 1932. He brought the power and durability of a V8 to the masses, just as he had the automobile two decades earlier. Previous to the Ford flathead any design with more than four cylinders was reserved for the ultrawealthy. Much of the reason for this was that the larger motors required complicated engine blocks that were cast in several pieces.

This then led to multiple difficult machining processes and delicate assembly to these high-stress components. If an engine block assembly fails at speed it's gonna be an experience. Something that could be inserted into the first 15 minutes of *Saving Private Ryan* but on the highway. A four cylinder motor could be cast in a single block.

Ford believed he could push casting technology to the limit and cast a V8 block in a single piece. That was the only way to bring the cost of a V8 down to the budget of the working class. How he pulled it off makes for a great story. He had secretly competing teams and they had different design requirements. He was a mastermind at getting the performance he wanted from people. He was also ruthless and never let a date slip.

The flathead V8 was unbelievably made at the cost point Ford demanded, the primary objective, and it was ready on the date he required, the second objective. But it did come at a cost. The motor was a reliability disaster. Ford was at risk of tarnishing a brand that was built on providing customers a product that was no frills, inexpensive, and reliable. It could have been very bad, except for one factor, he was living before the information age.

For Ford that was the deal to the customer: no frills, inexpensive, and reliable. This is how the product factors were balanced. People love their brands. Once they associate a brand with a certain factor balance it surely better not change, ever. It is too hard to disassociate or change what you expect from your favorite brand.

That's why companies create or acquire other brands when they want to enter new markets. The Lexus brand was created when Toyota wanted to market luxury cars. Toyota knew it would be too hard to have luxury car customers reframe what a Toyota was. Most of those power drills I was looking at in that Home Depot display were made by only a few companies that each represented at least three brands, each at a different cost point, feature set, and reliability.

Due to the speed of information travel in the 1930s, Ford had some slack in the risk of tarnishing his brand. Publications initially focused on the excitement of such an amazing product for the common man. After release, Ford worked to make improvements and get them to market before too much grumbling made it into the media or around the watering holes.

Over the next five years, they got all the bugs worked out. You read that correctly – five years – that's some slack. By comparison, the appliance manufacturer I was working with in the 2010s used the following mechanism to provide feedback to the team, the day after the new product release they reviewed all the online ratings from initial purchases. These reviews translated into action items that were delivered to the engineers mere hours after the customer angrily typed them in on their smartphone while on the toilet.

In summary, the allowable period of recovery for poor reliability for a 1930s Flathead V8 was five years. For modern-day home appliances . . . less than one day.

"No more free passes." If you have one saved from earlier in your career please turn it in, they're being recalled.

Understand the Intent

If we can't clearly describe *why* we're doing something, we should stop, agreed? It's generally good practice on occasion to stop and take a moment to reflect, "Is this worthwhile?" Whatever it is, making something, planning something, driving somewhere. Even if it was a great idea when we started, things can change along the way.

Another way to ask the "why" question is to ask, "What was the intent?"

During a program, I like to keep "intent" on our dashboard. Not at a general high level intent like "mission objective." A list of specific intents for all activity. "What's the intent of this Highly Accelerated Life Test (HALT)?" "Why are we doing this Failure Mode Effects Analysis (FMEA)?" "Why do I keep coming to meetings that could have been conducted through an email?"

I use "intent" so much that colleagues and customers refer to it as Adam's "I word." It's a powerful word, because it asks, "Is this still worth it?" Once we're in that frame of mind, we're estimating ROI. Keeping ROI estimations active on our dashboard means we can cut initiatives that aren't generating value and reinforce those that are.

Both cutting and reinforcing are powerful steps. Obviously, having actual ROI estimations to justify requests for additional resources is a smart way to go about project management. More importantly, cutting initiatives that we campaigned for, because things have changed and the value is no longer there, generates tremendous credibility. That's credibility that can be pulled out of the bank at a later date when there is a request for more resources in a high-value area.

The HALT test has been pushed out by six weeks. Should we still do it? The classic answer is "Yes." Why yes? Because if we cancel it then in the next program they will say we proved it wasn't needed. But the correct answer is to evaluate whether it still fits its original intent. Does it deliver the needed design robustness input at a time when the design can still change?

If it is too late to change the product then the intent is no longer valid and that money should be reallocated to another initiative, reliability or not.

Ever work on a grant-based project? If you have then you know how there is always a scramble at the end of the year to spend the rest of the budget. Why? So we can justify the same or greater budget for next year. Kinda backward.

Wouldn't you have greater faith in a leader who returned money she didn't spend? You might even reward her. Be wise in control of spending. Be the project manager that redirects money, so it can be spent where it will have the greatest impact.

How is this done? By evaluating a HALT test's intent, and deciding it can no longer deliver the value intended. With a late delivery on results, the test now has little value. Why waste program resources on it if satisfying the intent is missing? Parading around late test results just for the sake of it only hurts credibility.

With a HALT test, its value is to provide design input that improves robustness. Remember these design improvements won't be applied late. At some point, the design is locked down unless you find a real showstopper like "Kills the user."

Evaluating the value, purpose, and intent of the HALT test later than originally planned draws the conclusion of simply cancelling it and using that resource and time for another task. If that task is a nonreliability initiative you just may have made a friend who will help you out at a later date.

In Chapter 8 we explore how to ensure "intent" stays in the field of view by using the anchoring methodology.

Levels of Awareness

Awareness is everything. It's the first element of control. How important is awareness?

I'd say it's so important that I'd prefer to be in a bad place and know why I am in it than be in a good place and not know why.

If I'm in a bad place and know why, I can take actions to improve the situation. If I'm in a good place and don't know why, it leaves me vulnerable to change over which I have little control.

I'm sure you've seen a situation that degraded quickly because the people involved didn't understand why things were going well. (When I used to race cars, it seemed like every crash story I heard began with the phrase, "I was feeling pretty good, then. . .") If you were feeling good and then crashed, I'm pretty sure you were missing something.

For a reliability culture to work, then, people have to be open and honest. They have to be willing to share. When I begin an engagement with a new team or organization, I feel a sense of relief when the leadership describes not just their reliability strengths but their deficiencies as well. I'm confident that this type of team will grow to a high standard of product development.

What are some of the phrases I look for to see if they are aware?

- "We do/don't have good control of our field data, because of. . ."
- "We aren't able to differentiate between true design-based failures and ambiguous customer complaints or misuse."
- "Our products release without having completed reliability testing and we don't know how to make that happen. The programs move too fast."

When we hear statements like these, we're working within a team that can create a strategy that can be implemented.

I'm wary of projects that are under leaders who make statements such as:

- "We know what the problems are. We just need someone else to prove it."
- "We don't have time to do full reliability programs. Is there a test that can say it is reliable?"
- "Do you do 'certificates of reliability' for products?"

All of the above statements imply that the leaders are blindly confident. They think there are no significant problems. Or if there are problems, they can be understood by some shoot-from-the-hip analysis. These are the teams that are baffled at how they got sideswiped with all these "rogue" issues that came from nowhere. "I was feeling pretty good, then. . ."

A company actually asked me that last question: "Do you do reliability certificates?" I was like, "Yeah, here." I took a Sharpie to a sticker, wrote RELIABILITY across it, and handed it to him saying, "Stick this on your shirt." Then I left. OK, I didn't do the sticker part, but I did leave.

Just go ahead and make your own sticker, many companies do.

- "Trail Rated"
- "Field Proven"
- "#1 in Quality"
- "5 star reviews on websites"

A way to know if you're in this environment is that, when things go wrong, you'll find a general air of "How could this have been predicted?" There will also be names for problems that imply no fault, such as "gremlins," that's a bad culture.

Don't bang your head against the wall too much if this is the language surrounding you. Find like-minded individuals and see if you can persuade a change. If not, find a way to move away from the source of this mindset. It might mean going to another organization. If you're a leader and keep losing people who have first protested about why problems aren't being addressed, wake up before it's too late.

Summary

Intent and awareness are the two keys to success. Both are easy to have, and just as easily to lose. Like your keys and you phone. Grab them in the morning and you have them for the day. Or do you? Not if you don't keep track of them. We all do the pocket pat as we leave a room for a reason. Without that extra step it's easy to find yourself missing one or both by lunch.

The difference is keeping their status (keys and phone or Intent and awareness) on a dashboard, and looking at the dashboard (or touching it in the phone/key analogy). What would the inclusion of an intent and awareness status look like for your dashboard? Would it be something new or simply including them on one you already use?

References

1. Liker, J.K. (2004). *The Toyota Way: Fourteen Management Principles from the World's Greatest Manufacturer*. New York: McGraw-Hill.
2. Gladwell, M. (2008). Chapter 8. In: *Outliers: The Story of Success*. London: Penguin.

4

Awakening

The Stages to Mature Product Development

Stage 1

This kind of reliability testing usually goes under the name "verification and validation testing." It tests if the design does what it is supposed to do. Here, variability isn't much considered. Unfortunately, variability is what reliability is all about.

The intent of these Stage 1 tests isn't about learning. The intent is about passing. Of course, just trying to pass a test doesn't improve the product, almost the opposite. When we're trying to pass a test, we do everything we can to pass. We look at the easiest possible conditions, so things go well. What can we learn? Nothing.

During this stage, the organization often experiences large field failure surges. The pain of this experience – in terms of dollars, market image, and lost resources – awakens them to the benefits of incorporating reliability tools early.

Stage 2

In this stage, reliability tools and methods become a key part of the program. You may bring aboard a reliability engineer from the outside. Or, you could develop someone for the role internally.

Unfortunately, a significant number of planned reliability tasks get truncated or postponed when time and money get tight mid-program. Because of this, the impact of the reliability tools becomes reduced.

The company still experiences unexpected field failures regularly.

It's a difficult stage to push through, because they're investing in reliability and not getting much back. It's akin to someone who has started to exercise for the first time. The effort seems high and the results are low. This is when most people quit.

In this stage, there are indicators that show improved product reliability – things like fewer issues late in the design process and easier transitioning to a manufacturable product.

It's like standing on the scale and seeing you've dropped a few pounds even though you don't yet look buff.

This kind of encouragement is what gets the team to Stage 3.

Reliability Culture: How Leaders Build Organizations that Create Reliable Products, First Edition.
Adam P. Bahret
© 2021 John Wiley & Sons Ltd. Published 2021 by John Wiley & Sons Ltd.

Stage 3

Here's where the importance of reliability comes into clear focus. It brings together how the reliability program fits the product program and the company's overarching business goals.

What happens in the field? We see products that perform better than any previous generation.

We're now seeing clear muscle-tone and the return is matching our investment.

Stage 4

In this final stage, reliability becomes fully integrated into the program process and culture of the teams. An efficiency emerges that makes the reliability activities effortless while requiring little added investment.

We now have solid datasets. This helps analysis models drive design decisions. This is design input without testing, which is very powerful.

The test methods and tools needed in development are already available and do not require the expense of added time and cost to integrate. This is similar to when product development programs create in-house services, like machine shops and prototyping labs, which were previously contracted. The service is now right down the hall ready to serve the program. No red tape, no waiting.

Other departments are familiar with the inputs and outputs to reliability. Because of this, there is fluidity of information flow between team members and departments.

The Ownership Chart

Accountability is where the buck stops, and the teammate who's accountable is the one who makes whatever is supposed to happen happen. They're also the one who has to explain "why," when what needs to happen doesn't happen.

But, in an organization, knowing who's accountable for what can be hard. I'm not implying that people are dodging their responsibilities. Who's responsible for what simply gets confusing when the program is in high gear.

So if we're confused about who owns what, how are we supposed to track what's happening and who to go to when it's not happening?

Why do we want to address this in regard to reliability?

Well, reliability is one of those tasks that, when done right, touches almost all roles in a product development program. Which can also mean it is not easy to pin down the ownership for these tasks.

It's easy to say that the invention of the next advancement in laser tube technology is the responsibility of the laser R&D group. But remember: with reliability it's different. Reliability is being done correctly when the entire team is responsible for reliability. So who's supposed to be doing what?

We are all familiar with organizational charts. We often call them "org charts." Org charts suggest that they capture accountability and role relationships by showing us how roles relate to each other. Those of us who are seasoned know that the org chart does not capture accountability or information flow as it actually happens, at all.

I propose a different type of chart to supplement the org chart. It's called the "ownership chart." The ownership chart came about because I frequently found myself sketching diagrams of who in the organization was taking responsibility for reliability activities and who needed to be. I needed to diagram how information flowed and who wanted it. Sometimes I even found there was no internal customer for information that was being published regularly.

Comparing Charts

Soon enough I was sharing these sketches with executives to help them understand how things were currently operating, and more importantly how I thought they should be operating. Next thing I knew, these charts were a standard part of my assessment reports.

Simply put, these charts bring the intention of an organizational structure into focus. (There's that word again, "intention.") In other words, when it comes to the reliability process, what is Jennifer accountable for? How do her accountabilities affect Sam? What information flows between them to support these accountabilities?

A standard org chart doesn't have much to do with accountability. Its main job is to diagram the interactions of the team, based on managerial needs and hierarchy.

An ownership chart, however, is all about accountability. That's its sole function. The chart identifies who generates certain types of information, who makes specific decisions, and who's able to direct, block, or pass all this information.

If an individual can create information, they need to be identified as an owner. A test technician's lab generates information. In a traditional organization chart or role description, this technician may not be considered accountable for test results. But what happens in reality?

I've seen test technicians report information directly to middle management. The information they were generating was so critical to the program's day-to-day operation that they, the test technician, would attend the Friday program steering meeting, hosted by the CEO.

This critical lifeline of information wasn't documented. You couldn't find it in any program plan. Unfortunately, undocumented means unprotected.

Ever wonder why a new program can have a rough start when the previous program, using the same team, was a well-oiled machine?

It's because things that were happening to generate meaningful results, like a technician reporting information to a CEO, were never documented. A small, undocumented practice like that likely won't continue in the next program. That's because to write it down seems ridiculous. Those involved may not realize its importance. Or, they may be embarrassed that a CEO had to go to a technician to get critical information.

Benefits of the Ownership Chart

Once an ownership chart is created, it makes it crystal clear how information gets lost, stuck, and sent in the wrong direction daily.

These are two responses I have heard after a team first views an ownership chart:

> "Why are those summary engineering reports going to the VP of QA? The R&D director uses that information more than anyone else."
>
> "It doesn't make sense that the system's engineering manager has to review the design risk summary at every stage. Let's have the team-lead be accountable for this information getting to the project manager. They can rebalance the project resources."

So how does this help reliability? Two ways.

The first is that reliability tests generate information with a shelf life. This "aging" information not landing in the correct hands at the correct time reduces its value dramatically.

The second is the reliability department isn't being informed of what the program needs. How can they provide good value if the needs aren't clear? Being aware of what is needed permits them to arrange tests and analysis that can yield information when it's needed.

Generating reliability predictions and robustness measurements takes months. If the request for reliability information is made when it is needed, it is already too late.

These are the steps to create an ownership chart.

Begin by creating nodes. What's a node? It's usually the people, positions, or groups involved in the project. They're the ones generating or delivering information.

Many of the nodes you'll be creating can redirect where the information is headed.

The information that the nodes will be handling can be either pushed or pulled. What's the difference?

In my household, both my wife and I could be considered informational nodes. When I get home from work, we'll likely exchange information. If I tell her that on my way home I saw a rare 1960s Ferrari, that's information I'm pushing on my wife. In other words, she didn't ask for it, but I'm telling it to her anyway. If she then asks me if I picked up the item she wanted from the store, she's pulling the information from me; it's information she needs to do the project she is trying to complete.

Here's how that works in the office: as a project manager, I may need the "reliability life testing" results to decide if we can move to the next phase. This is information I'm pulling. As a quality engineer, I may have found an increasing failure rate in the field for our flagship product. I'm going to find project leaders who I can share this important information with. This is information I am pushing.

Next, identify information that's critical to program decisions. This is what's flowing between the nodes. This may be field failure rate updates, projected failure rates from testing, or even a simple program resource change.

These questions will be helpful in extracting the information needed to create an ownership chart:

- Where in the program will information be created?
- What information do people need for their roles?
- What information gets pushed onto people regularly?
- What information does a role "touch," but which adds no value?
- Where is information needed for program resources and schedule decisions?
- What information is needed to make major phase gate decisions?

As you can imagine the ownership chart differentiates greatly from the actual org chart. Laying the two over each other highlights how different teams actually interact compared to how they are structured to interact. In Figure 4.1 we have a few roles in the organization with connections based on org chart roles. In this chart the blocks are roles and the connecting lines are direct relationships. A direct relationship is where information and requests flow.

Now let's erase those lines and add new lines that show accountability for pushed or pulled information that occurs on a regular basis (Figure 4.2).

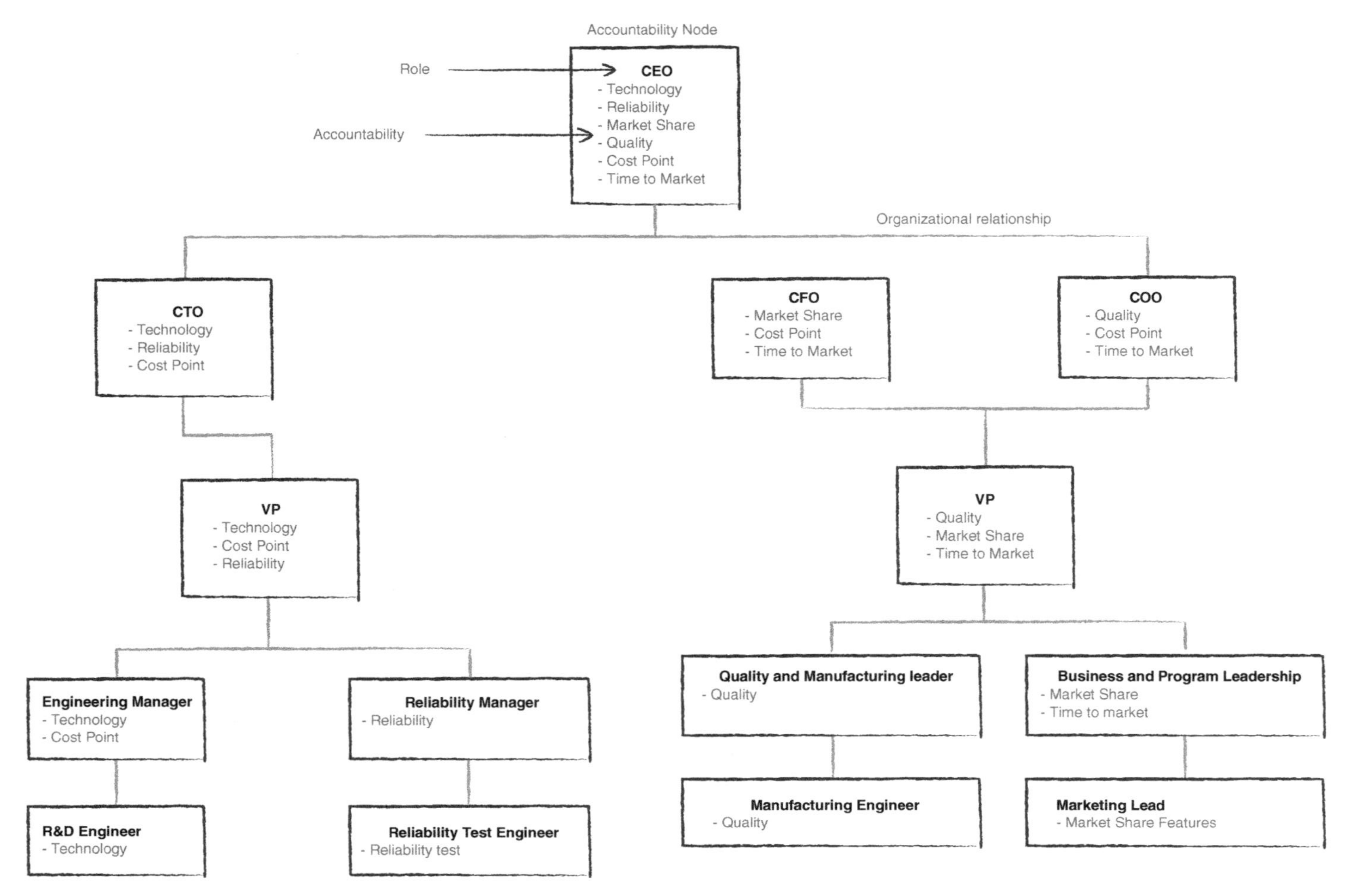

Figure 4.1 Ownership chart.

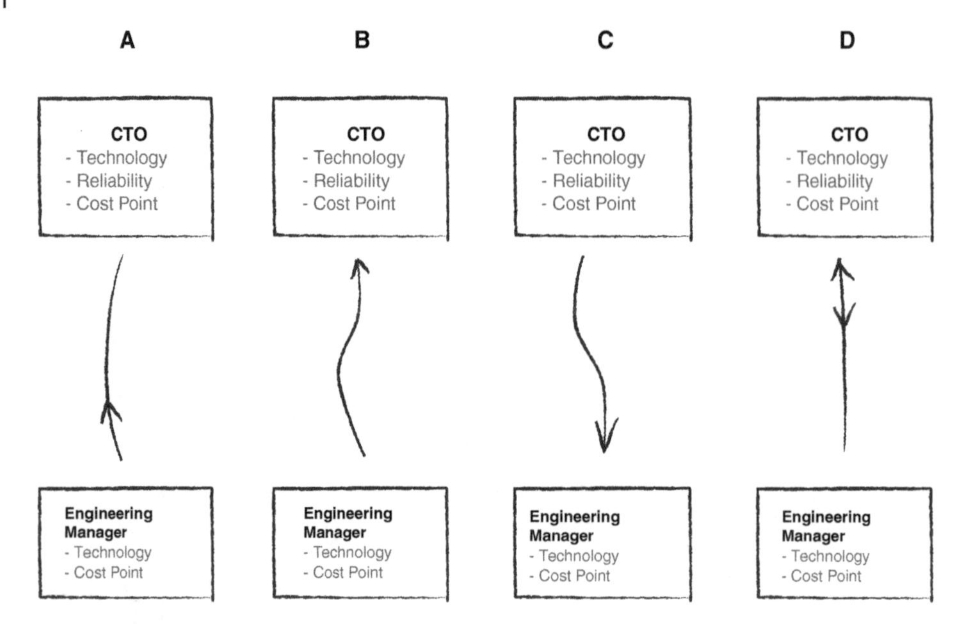

Figure 4.2 Accountability notation.

Our notation will be a drawn line with one or more arrowheads on it. The direction the arrowhead is pointing is the direction of the push or pull. The role the arrowhead is closest to is the individual that has initiated the interaction. What does the chart reveal when accountability is drawn? Just look at Figure 4.3. In example A the engineering manager is pushing information onto the chief technical officer (CTO). In B the CTO is requesting information from the engineering manager. In C the engineering manager is requesting information (pulling) from the CTO. In D the CTO is both asking for information and pushing information onto the engineering manager regularly.

Let's look at an accountability chart using the original org chart (Figure 4.3).

This spaghetti mess is from an actual evaluation I completed. Here is my summary.

- The CTO needs to stop both asking for updates and giving directives to the manufacturing engineer and reliability engineer (first level people).
- The R&D engineer needs to stop telling the CTO what is going on and the CTO needs to stop acting like he likes it by asking more questions.
- The CEO needs to stop asking everybody for updates on everything and get a dashboard or something.
- Why is a manufacturing engineer waltzing into the CEO's office to tell him what he did today?

The accountability for information is all over the place. Obviously, the roles as they are defined do not satisfy this program's needs. The information flow doesn't have to follow the org chart, but it should make sense. In this example, the second-level managers are out of the loop on a good deal of information on topics they are held accountable for. But you can also see that it doesn't look like they are asking for much if their subordinates are trying to find people in the chain of command to listen to them.

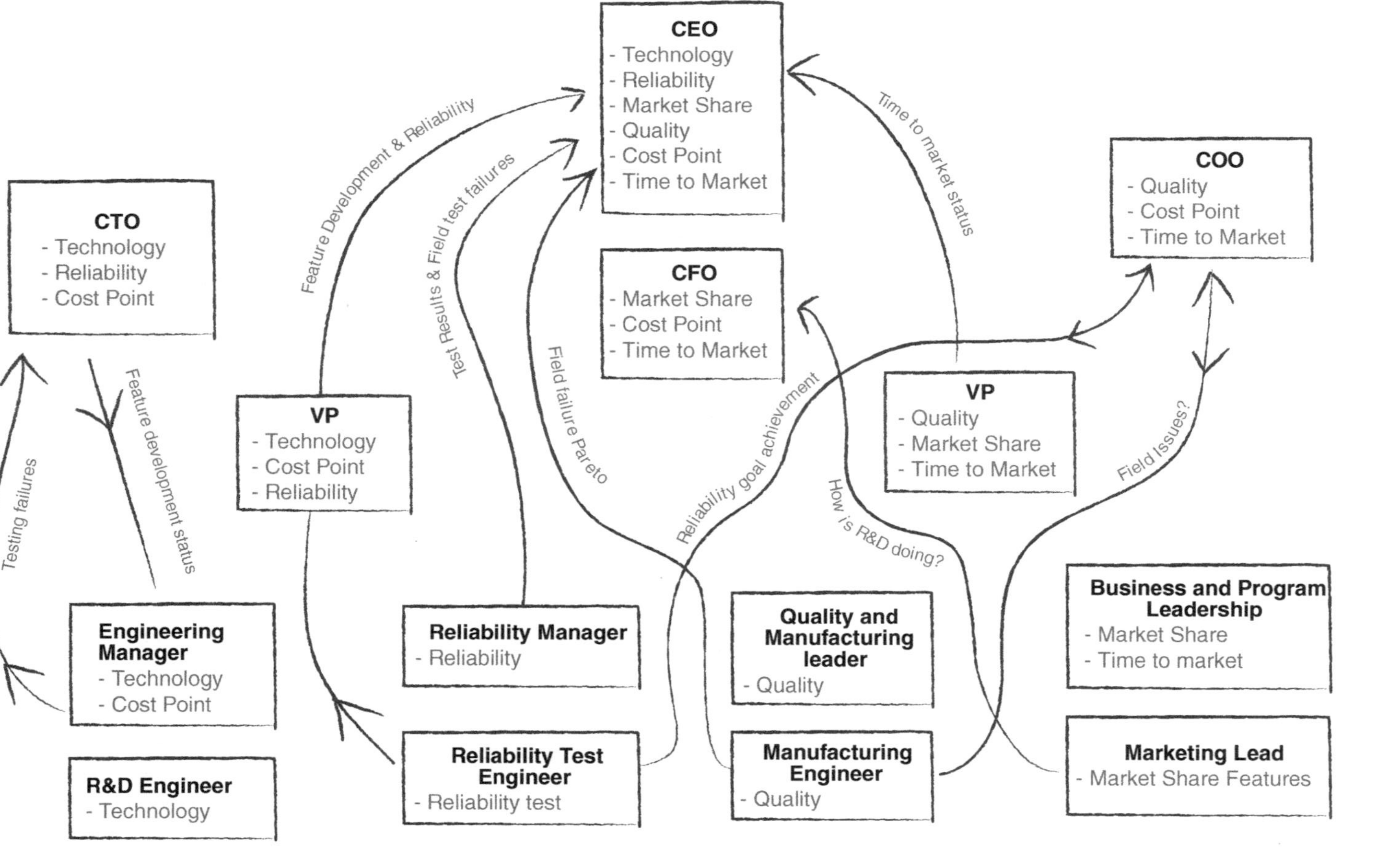

Figure 4.3 Accountability chart.

Here is an example of a specific issue that came clearly into focus once I created a path for accountability in an ownership chart. In a set of interviews with a team, I found first-level individual contributors were updating a VP on test results and asking for guidance.

In this organization, the executive leaders were overwhelmed by the number and variety of decisions that they had to make. It was amazing that they could lead at all. I wouldn't have been able to get out of bed in the morning, knowing what was ahead of me.

In this organization the product's reliability performance was entirely the responsibility of the technology team, not even the project managers. There was an unspoken guideline that "the product would be designed reliable." OK, but how was that defined?

> "Hey, Mitch, did you design that reliably?"
>
> "Yeah, did you do your part reliably?"
>
> "Yeah."
>
> "OK, good. I guess we're all set. Let's call it a day."

There was almost no accountability for the reliability goals between the development team and the VP. Every person between them believed that it wasn't their job, and they were right.

How could it be that the president cared about these reliability goals so much but the bulk of the organization handled them like afterthoughts, wishes, or desires? This ownership chart was going to do amazing things for them. Like having chronic chest pain and then finally seeing an x-ray showing half your ribs are broken.

Take a moment and diagram a path for critical information in your process. Not your planned path, the actual path. Ask a few questions, play detective for a day. It will likely become evident that information is not being handled well. Changes are easy after that.

The VP that was the receiver of regular frontline test updates was under visible pressure that was excruciating to witness. The VP's primary responsibility – to strategize and lead an organization – was on most days an afterthought.

When the ownership chart was created, what became immediately evident was that things would get much better if mid-level team members shared ownership of reliability.

I created an ownership diagram that accounted for all of this. It fascinated the team to see where ownership for initiatives like product reliability really lay, and how the owners were connected. How obvious were the solutions? We outlined about 60% of the new accountability paths in that first day of analysis. It was that clear what had to be done.

Communicating Clearly

In any organizational assessment, it's critical to listen closely to language. Big organization, small organization – it doesn't matter. The thing that moves information along is language, or dialogue.

Listen for the dialogue's intent (there's that "i" word again). The intent is not the content. Take a step farther back and listen behind the words.

When you're handed a task list, it often has an underlying theme. But you may not pick up on the theme right away. Simply completing the list's action items, without looking deeper into things, won't make the list owner satisfied. ("Did I really listen to what was asked?")

OK, let's do a little "listening behind the words." Here's a quiz.

In the following example, what's being communicated? What's the intent?

> "Go straight for two blocks. At the stop sign, make a left and go two more miles. On your right will be the store you were looking for. The building is white, with a red sign."

The intent is to provide information. Every word provides information concisely. Nothing extra was added.

Now, listen to these directions.

> "Go straight for two blocks. On the corner is a cafe with the best pastries. I don't know how long you're in town, but if you can get there some morning, I recommend you grab a croissant or even just a bagel. Anyway, take a left and go two more miles, and you'll see the store on your right. But, in case you're interested, I'd also look in Simon's, which is three blocks farther. Their selection is bigger."

With those directions, the intent isn't as clear. There's an interest in socializing. Giving directions was of secondary concern. Their clarity was sacrificed with so much peripheral information added.

The listener will miss important details. The value is diminished, because of the secondary objective. This flips the priorities around.

What if the direction giver answered this way? "Go straight for two blocks and turn left." They then abruptly walk away. I'd say they don't like strangers or are tired of all the tourists in town. They are getting rid of the listener.

Behind the Words at Work

What do you take away as the intent of the following paragraph? This is a word-for-word quote of dialogue in a meeting I attended.

Remember, listen behind the words.

> "We're concerned with the primary risks. We need to complete this DFMEA and the regulatory risk analysis, because if we don't catch all of our issues we are going to be in the same situation we were before. It's the risk analysis that's important. Without it we won't be able to understand what parts of the design we should be most concerned about."

Have you ever heard someone in a meeting talk like that? Too often, I'm sure. It's exhausting. When it happens to me, I have to make a conscious effort not to roll my eyes.

So what was behind the words in that design failure mode effects analysis (DFMEA) speech?

Nothing. It doesn't contain a single piece of information or directive.
I think this person is scared. Let's break down what they said.

- "We're concerned with the primary risks." Ahhh, yup! Of course we are. Risks are concerning. Zero information here.
- "We need to complete this DFMEA and the regulatory risk analysis, because if we don't catch all of our issues we are going to be in the same situation we were in before." That's totally true, and we knew it already.
- "It's the risk analysis that's important. Without it we won't be able to understand what parts of the design we should be most concerned about." That's correct. That's what a risk analysis does; we have dictionaries. ("Don't roll your eyes, Bahret. Nod and smile. He has no idea what you're thinking. You're doing great. And the Oscar goes to. . .")

Passing along information was never their intention. What was their objective? They used authoritative jargon like a protective shield. That's what a politician does. They're careful not to say anything that they could be held accountable for later.

My conclusion is that this person, who could have considerable knowledge, won't actually contribute, because they're scared. Scared of one of two things.

They're either operating in a culture in which those most likely to survive are "yes men" (an authoritative culture). Or, they may feel they're not a valued contributor, so their confidence is low. I've seen this in situations where an innovator is in a non-innovative process – a fish out of water.

I can determine which of the two reasons is at play through a facial tell. That tell is a smile.

If there's no smile, this person is likely concerned about management's imminent reprimands. They can reference their seriousness when issues arise. It works like a "Get out of jail free" card. It's an "I told you so."

If they do smile, they're concerned about not being a true contributor. They want you to listen to their important contribution, but they then want you to know that they're just trying to help, "I'm not looking for any trouble, just offering some wise counsel." They're effectively backing down from any challengers in advance with that smile.

The next clue to understanding the full situation is to look for repetition of this empty talk in the group. The repetition indicates if this is being driven by the person themselves or by the culture. The culture may be the result of a particular leadership style.

There are many examples of this leadership style outside of corporate life. Politics is a place where the "strongman" technique is often found. One example that was captured on tape is when Saddam Hussein made his move to take over his political party. The video seems like a scene from *Scarface* or *The Godfather*. It's not indicative of what happens in corporate dynamics, but it's easy to see the parallels.

In the Hussein video, there is an auditorium full of men. They're listening to speakers from their political party. Hussein approaches the podium and introduces an individual who'll speak next. It is a party member who's going to confess to plotting to overthrow Saddam Hussein. He then lists those working with him. As he recites a name, the corresponding individual is led out of the room by two armed guards.

While this happens, other party members stand. They declare their loyalty to the party and to Hussein. Saddam sits quietly, leans back, and puffs on a cigar. He just took control of the party.

Have you been in a meeting surrounded by yes men? "I agree." "Me, too." "I agree even more, and I'm going to restate what was said already, but with more conviction." You can almost smell the lit cigar.

If you witness multiple people talk in this manner there's a strongman in your midst. Scanning the field for people to make an example of. If you hear talk of this type and you're the leader, you may want to take a look at how you deliver your leadership. I've met leaders who have no idea this is what they've turned their product culture into.

When You Want to Improve

I follow the advice I give clients: "When you want to improve, hire someone who can guide you." As a consultant, this is more important because there isn't a colleague on the other side of the cubicle wall to bounce things off of. Here is a semi-relevant story.

When my oldest daughter Katie was eight, she set up an advice booth in her room. It was our old hand puppet theater. She charged 25 cents a session. We considered the price reasonable, and she had a solid clientele.

My sister, Aunt Abigail, was visiting. Katie proposed that she and Abigail conduct a session. Payment was required up front, cash only. Abigail figured the cuteness of the whole enterprise was worth the quarter.

She, however, came away surprised. She actually got some great advice. How could an eight-year-old know that the best way to make a big career decision, involving a transfer several states away, should be driven by "Take a quiet moment to think about what will really make you happy, and don't be afraid"? Abigail could have very well paid $200 to a licensed therapist and received the same advice. She got the deal of the century.

So, how could this be? How was it that someone who'd never been in that situation or for that matter had to make any sort of big life-changing decision proffer such sound advice?

It's simple. She didn't have the answer to the complex question. She was simply telling everyone the same piece of wisdom over and over again. Maybe she heard it from some cartoon character on TV, or from a teacher. What made it valuable was that she was on the outside, not consumed by the issue.

A lot of problems can be solved by walking around to the other side of the puppet theater. But that can be hard to do. So it's always worth the quarter to get an observer's perspective.

My Personal Case

I have hired many consultants while being a consultant: writing consultants, graphic art and web consultants, mathematical model consultants, market positioning consultants, and business strategy consultants.

A single moment of growth surprised me more than the rest. It came from a business strategy consultant. He told me, "Your biggest growth problem is how you speak." It turned out I wasn't such a good talker.

The consultant told me there was no reason for us to continue with our business strategy work if I didn't improve how I spoke to leaders. He was going to quit on me if I didn't fix that first.

He was right. I had enlisted his services to improve making direct connections with senior executives. I figured what was needed was to simply create a message that would help executives "get" these principles. But that was backwards. They didn't need help listening. I needed help explaining.

My consulting company began as an engineering firm solely. I sold to and worked with engineers. Easy-peasy, I can talk to engineers all day long. Hey, as I'm writing this, I'm wearing a t-shirt that reads, "MAY THE d/dt(mv) BE WITH YOU." Get it? You might not, but my people would. This is how I pull them out of the crowd.

I spoke the engineers' language. I knew what they were thinking. I was them and they were me.

On the other hand, executives weren't "my people." I was never a CEO or had managed a thousand-person organization. I didn't know what they were thinking. But that didn't mean I didn't have the information they needed. Katie's success was based on beautifully articulating the message. She knew she had to speak "adult."

Getting the Message Across

The biggest hurdle to getting my message out was how I spoke.

What I learned concisely was this: when engineers speak and write, they tend to cover all the bases and offer every bit of information possible. They heap on facts, figures, references, and probabilities.

Picture someone going to a library help desk and asking for a reference on algebra. If the librarian was an engineer, they'd go to the stacks and come back with a dozen fat algebra textbooks. Then, they'd top it off with some academic papers on how to learn math. With all this information, the person who had asked for help would likely just walk out. That's what happens to engineers in conversation, too, and it makes us pretty confused. After all, we're only trying to be as thorough as possible.

Thoroughness, by the way, isn't always a drawback. Thoroughness is a quality you want in the guys who design your car's seat belts, airbag, and brake system.

But our thoroughness is why we get the reputation for being condescending.

Talking down to people is rarely our intention. We're just making sure that the listener has enough information to build a Mars habitat from Mars materials. Even though the actual question was "Do you think people will live on Mars someday?" It's easy to pick out an engineer at a non-engineer social event. Just look for all the rolling eyes around them.

The Importance of Time

Time is an executive's most valuable commodity. They're smart, confident people. When speaking with an executive, the most important factor is to share only the information that's most necessary. They'll ask questions if they need to know more. They're not timid or worried about how their question makes you feel. You don't become CEO by being worried if everyone's having a swell time.

What I did consistently when I spoke to leaders was classic engineering-talk . . . with a dash of insecurity on top. I wanted these executives to like me and I needed them to know I knew a bunch of smart stuff. If I was the guy giving directions in that earlier example I was thinking, "Wow, they're going to think I know lots about this town and I'd be fun to meet for breakfast." That wasn't the reality. If they wanted a breakfast date they'd ask.

Looking back, it's obvious how many conversations with leaders I've blown. With the knowledge I have now, I can identify the exact moment I blew the conversation. I can pick out the exact point where many executives went from interested in what I had to say to working out a way to end the conversation.

One such conversation began with a call I didn't expect. This CEO was reaching out to me as a cold call. We'd never spoken previously, but he wanted to start a collaboration based on what he had heard and seen of me.

I blew this call badly. I spoke over him, gave answers that rambled, and told a stupid story. This was all stopped by him saying, "Well, I'm sure you are busy, so maybe we can pick this up another time." Ouch!

But I learned something that's applicable to the reliability leadership methods in this personal growth moment. Dissect language and define it's true intent.

That CEO quickly realized I was desperate for something, not his respect but his affection. This was likely going to get in the way with the task at hand. He also suspected it might be hard to work with me because I couldn't be concise.

When We Can't Communicate at the Organizational Level

Simply use the intent of language to dissect how an organization actually operates. It's right there in front of you. This is an example where understanding the intent could have saved a disaster.

Unfortunately, this is a real story, and I was in it.

I was consulting with a company that was developing a cutting-edge piece of technology. The market situation was that a competitor had released a new product that was a big jump, compared to what my client offered. My client now had to attempt to displace the competitor as the leader. My client was not a startup. They were a mature, decades-old multibillion-dollar company, with large divisions in six countries. They knew how to develop products.

Leadership felt the most critical element to the program's success was time to market. This advancement in technology was going to have to happen at a dizzying pace. But the mistake leadership made was simple. They would undermine time to market by focusing on it singularly. They made getting the product to market as quickly as possible the team's be all and end all.

They sent a clear message to the team: "If you get in the way of the project's release date, prepare yourself to be run over." I don't think it was ever verbalized that way, but actions, reactions, and directives made this message clear.

Leadership knew reliability was a critical element to success as well. Customers don't see value in new technology that can't be depended on. This was not a consumer product. This was a healthcare product and a failure in the field could have serious consequences. Leadership thought that a reliability program was something that could just be bought and wouldn't disrupt the program timeline. Engaging me as a consultant was a part of this strategy. I had free reign over the team and a solid budget.

My Own Experience

My job was to ensure the best reliability practices were in use. I was operating at about the equivalent of a director. This engagement was initiated by the senior VP who had sought me out personally. But in the end, *I* should have been advising *him*. Directing the team wasn't enough for this program to succeed. We were not operating in the best program culture. I think today he would agree with this statement.

I was directly engaged with multiple engineering teams. They were in Massachusetts, New York, Utah, and France. Only the New York and Massachusetts teams had worked together previously. The team in France was part of a recent European acquisition. An added challenge: they were not only in a different geographical culture but they had a business culture that was different from the rest.

The Utah team was a separate division that, for the most part, hadn't engaged with other divisions before. They in fact had excelled without much input from the outside. Having to work with these other divisions, then, bothered them. They responded with a bit of hostility.

If you've ever been to Utah, such a response is a bit surprising. They're mostly outdoorsy people who love nature and talk about the natural aura connecting us all. At least that's who I engage with when I go to the Canyons. These people I was working with didn't seem like Canyon hikers. They gave off more of an NYC taxi driver vibe.

The disdain from this group was so palatable that after my first conference call with them the Massachusetts (Boston) team apologized for the Utah team's behavior. Stop for a moment and think about that. These are Boston people thinking that the other team was rude. Have you ever been to Boston? We punch people in the face for messing up our coffee order.

I was the first person brought in to begin the reliability initiative for this project. The organization had no precedent for this type of project. They had a reliability department, but they had only focused on less complex products. For them, this was new territory. I had experience with a similar type of technology in reliability programs.

I started the engagement by having them do a few direct hires to build the group up. Some were dedicated reliability/design engineers, a few technicians, and a reliability project manager to be the day-to-day leader. By about a year all of those people, except for one technician, had left or been asked to leave. The project manager had a meltdown and was asked to leave shortly after. It was a "show results or you're out" kind of atmosphere. I don't know why they kept me and the technician on. But we did have a whole office area to ourselves, so that was nice.

After the purge/exodus, I built the team up again. I pulled in some day-to-day leadership from another program and a few new hires. The technician that stayed eventually ended up leaving on his own accord as well. He had had enough. He said he just couldn't stand working with his hands tied. I get it, the ropes chafing me, too.

So I was the only person who was consistently part of the reliability initiative from the beginning. When the reliability program continuity is based on an individual who technically doesn't even work at the company, that's a big red flag.

Why Testing Doesn't Happen

It was almost impossible to get all the testing done that we had planned. Engineers were constantly told their design and manufacturing deliverables were higher priority, putting reliability on the back burner. Even if the team had been more available, testing was impossible because the designs were repeatedly being scrapped. We didn't have any recent design revisions to test, so most of the results were meaningless.

Then we had our regulatory-driven requirements, which forced design freezes. Any improvements we recommended for reliability robustness might as well have been thrown in the trash after a design freeze had passed. In many cases we started testing after the freezes occurred, so the advised improvements could not be included. It went on and on like this through the program.

For most of the program the one thing the executive VP said more than anything was, "Why do I walk around the facilities in Massachusetts, New York, Utah, and France, and I don't see any tests running?" He would say this again and again for two years. He would get roundabout answers from everyone that sounded like a pre-teen explaining why they didn't take out the trash. "I have homework." "I thought Sara was going to do it." "I'll do it tonight."

Those conversations got to the point that I just stayed out of them. The five most common reasons as to why tests didn't start or continue were:

- The design doesn't work and in this phase there won't be a redesign.
- The design engineers had been told to work on tasks other than reliability.
- There was no money allocated for units to test.
- The contract manufacturer making the test fixture was late.
- The safety team stopped us and wants a redesign of the test fixture.

But at the end of the day the VP knew that these were all localized reasons (excuses) for specific instances. He wanted to know why this was occurring across the board. It couldn't be that there weren't units to test, no engineers with bandwidth to participate, no designs working well enough to be tested. What was it?

It wasn't until about three-quarters of the way through the project that I realized what the reason was. Why the VP could walk around every facility and see almost zero tests being run. The issue here was that the pressure from above was shaping what "failures in test" meant: "engineering failure." Failures weren't valuable information to assist with improving the design, as they should be. Failures were a spotlight with a flashing neon sign that stated "You are holding up the program!"

This was why there were almost no tests being executed when systems were available. As a designer or engineering manager, the perception of a failure in a test was a serious deterrent. Why start an activity, testing, that almost certainly was going to result in your having a target painted on your back?

When Scheduling Trumps Testing

The logic was simple. All test results were immediately reported to management and all tests of new designs had failures. Failures slow down the program. Slowing the program was the worst thing that could be done in management's eyes. So testing your design was a bad idea.

The leadership had transformed the culture into what resembled a freight train with a nonnegotiable arrival schedule. I dare you to get in front of that. That's what testing was.

Leadership was obsessed with finding out what was causing delays and removing the obstruction. The drive to get to market quickly had been such a dominant factor in the program's creation that it was poisoning all other aspects of good product development.

OK, so we all have been in rough environments. There must have been some engineers or managers with enough clout to do some testing and survive taking a hit? But why would they? There was no time to make any design changes with the pace the program was going. Testing was a "nothing to gain, everything to lose" proposition for even them.

The result of all this was that the program would use the first release field performance as the testing to fix what didn't work. By the way, this was a surgical device. A surgical device! We're going to use first release field failures as feedback for design robustness improvement?

The guy lying on the surgery table is your test fixture and you just demoted a surgeon to a test technician. This is our plan because you think time to market trumps everything else?

When laid out in black and white, this all sounds insane, but I know that more than half of you reading this book are nodding your heads. You've been in that environment before, haven't you? It's something you slip into slowly, like a drinking problem. This had become a culture of fear.

At some point I had enough and told them I was leaving. I just couldn't stay a part of this and consider myself a good contributor in product development.

After disengaging, I met with the executive VP for coffee a few times, just to catch up. I wanted to hear from him how things were going – a bit of a candid debrief. There wasn't much to hide since the news was out about the delays. Wall Street had publicly chastised them and nobody was pretending things were going to be OK. The environment back at the office was now a bit more, "What happened?"

He did bring up how he still didn't understand why he couldn't get anyone to do testing. I shared why I thought this was.

I recall very clearly the way his face changed as he realized it was the fear he and his colleagues had created regarding time to market. This is what drove the team to avoid testing. He drove the team to avoid testing.

When that clicked for him, really clicked, there was a moment of silent pondering. He composed himself to the neutral executive expression we see in meetings and said, "Hmm, maybe you're right." That was the last we spoke of it.

But hopefully that realization led to some different cultural dynamics in their organization. Probably not. It would have taken not just his realizing this for there to have been change. He was also reacting to pressures put on him.

A year later I saw an article in a major magazine that said the release of the company's product was delayed. The irony that they couldn't get to market at all because the design was so underdeveloped was not lost on me. It made me sad because I had seen the talent and the amazing technology, and knew what could have been. I had failed, they had failed, everyone had failed at creating something that could have been great and helped patients.

Summary

Awakening is not an absolute. It is a term that describes a comparison between a past state and a new one. Before, we had a level of observation of our surroundings we considered representative of reality. But what we can see changes. Eventually, our perspective of what is real is different.

This is just part of the human condition. We can all look past to 1 year ago, 2 years ago, or 10 years ago and see how we have progressed in wisdom, awareness, and understanding. What I hope you take away from this chapter is that with the action of measuring the degree of awakening we can track progress and then change what we are doing to keep it at a steady or even faster pace.

Side note (I can't resist): would tracking the rate of progress of awakening be Awakening: d(awareness)/dt. So a team's total value(score) of awakening would be

$$Awakening = \int_{Program\ Start}^{Today} awarness(\#\ program\ metrics)\frac{d}{dt}$$

Calculus joke, but I actually think that could be used in comparing if a team is improving the improvement process. So if the integral of your current awareness growth is greater than your past period of awareness growth, you are "improving" your improvement process. I say we slap that one on the program dashboard. Hmm. . . look for that in the second edition.

5

Goals and Intentions

Testing Intent

In many program cases I see teams "testing to pass" when they should be "testing to improve." Testing to pass is putting your best foot forward. Put on your best suit or dress, comb your hair, smile big, and always give a firm handshake. There is a "mark" and you are going to hit it so you can advance to the next stage.

Testing to Improve

Testing to improve is looking for defects and "performance response to variability." By "performance response" I mean failures. Let's not sugar coat it. These failures occurring in the team's hands make up a key element to improving the design.

It's more akin to going to the doctor's office and wearing that gown that is very convenient for examination. Not taking off your street clothes in that situation is just wasting everyone's time and makes the whole learning process take a lot longer. That is what I see many design teams do. Test the perfectly hand-built units, use nominal stress, and throw out data that indicates issues have occurred, often calling it an "outlier." You just threw out gold! That failure is what we were hoping to find. The customer is going to find it next.

Quick Question

If the program pressure is to release on time and your personal annual review is right after product release would you. . .

A) Explore all the design's deficiencies?

or

B) Make sure it passes each test with flying colors, not being accountable for delays in product release?

OK, now let's say you are designing a new plane and your family is going to be on the maiden flight. Would you?

Reliability Culture: How Leaders Build Organizations that Create Reliable Products, First Edition.
Adam P. Bahret
© 2021 John Wiley & Sons Ltd. Published 2021 by John Wiley & Sons Ltd.

A) explore all the design's deficiencies?

or

B) Make sure it passes each test with flying colors, not being accountable for delays in product release?

That's the difference between "test to pass" and "test to learn." The challenge is for leadership to manage their teams to be owners of the design, the same as they are. If the team is testing to pass, it is usually due to no error of their own.

A team's personal success or even household income depends on it. The difference between most product development programs and the airplane example is their family isn't on the plane. Their family just needs that paycheck to pay the mortgage.

I originally created the "family on a maiden flight of a new airplane" example to demonstrate the "test to pass" principle many years ago. But in a way that example became real life.

In 2019, there were two Boeing 737 crashes that killed all on board. The 737 was a newly released design. If you are reading this far in the future and are unaware what happened, this is what happened. The root cause is simply that Boeing pushed product development programs so hard with "time to market" goals that poor and untested design was going out the door. At the end of the day all the analysis on software and system errors boils down to one simple statement, "Boeing let market pressure drive bad engineering practices."

These product time-to-market pressures grew to become such a gargantuan force in development programs that not even a situation where families' lives being at risk was able to counter it. We need to change how we balance our product goals through the product development process now more than ever.

It sounds ridiculous to suggest the words "fear" and "ownership" could ever be mistaken or considered equivalent. But so many leadership staff seem to act as if they are. When they get all "tough guy" and demand a subordinate accomplish a task, are they transferring ownership or just creating fear? It's just fear, fear of being held accountable, fear of expressing needs to their leaders to accomplish the goal, focusing on collecting evidence of no wrongdoing to defend themselves in an upcoming trial when things go wrong.

Ownership

Ownership is different. Ownership of an outcome results in an individual doing all that's possible to ensure the result is achieved. Uneducated leaders think the fear route is a quicker path to the desired result. It's not.

I've seen this dynamic drive a team's testing. They test and collect evidence solely as a means of covering their backsides when things go wrong. That's messed up!

If the company owns the design and ultimately its performance, why hasn't that same ownership been transferred to the design team? Simple: it wasn't transferred. Laziness, uneducated leadership, doing it the way we have always done it. It really doesn't matter why. Rice farmers don't act this way; they can't.

Fear-driven Testing

There are some hallmarks of fear-driven reliability initiatives that are easy to spot. A classic indicator of fear-driven testing is, "We need to do vibration/temperature/humidity testing."

Why? Why do you need to do something so nondescript? To say you did it. They're just checks in a box you can later point to: "We did vibration testing." It's the equivalent when business strategies are based on "synergy," "collaborative style," "crowd sourcing," or "interconnectivity." It's empty.

Leading a department earlier in my career my colleagues and I were submerged in a culture of empty initiatives. I knew we had totally given up when we started playing "Bullsh** Bingo." This is how we played it.

Each of us would pick five popular buzzwords we expected to hear in the next meeting. If all five of your words were mentioned, you won that round. The game had two additional rules: you weren't allowed to say the words yourself and you couldn't lead someone into saying them.

If you're a part of a leadership strategy initiative, ask yourself the following: "Can we connect the initiatives to a clear intent that delivers a product or program goal?"

If this connection isn't clear then it should be pointed out immediately, and addressed. Here is a way I direct a team to re-evaluate what they're doing when a test initiative is laid out and I suspect we aren't doing it for the right reasons.

TEAM: Should we do vibration testing?

ME: I can't answer that without knowing why we'd do it. Let me ask you a few questions to get us there.

- Do you want to know how much margin the product has on an expected vibration level?
- Do you want to know what fractured parts from poor manufacturing look like with a small sample size?
- Do you want to know what the random failure rate is for the product during midlife?
- Do you want to know what the time to failure from accumulated life stress is in use case #2?
- Do you want to know about the product's performance or design?

Until we start at the beginning and identify what you want to know, we're doing testing for testing's sake. That's called a "phantom testing program."

Transferring Ownership

Peter Drucker said, "Leadership and management are entirely different concepts: Leadership is doing the right things. Management is doing things right" [1].

If we want to become leaders, then, we have to do the right things. And to do the right things, we must take ownership.

A funny thing, though? As leaders, we often must transfer ownership to the people closest to the process. The reason? We're too far away from the day-to-day workings, and we can't provide enough guidance. But the people working closest to the process can.

In transferring ownership, we can't make assumptions and keep important directives to ourselves. We have to make it clear that nothing matters if the program goals aren't met. Game over for the product and the success of the business.

There is no trial or inquiry after the failure. Simply, they must make it happen, not because of perceived consequences but because of real consequences.

Leadership and Transference

As I described in the previous section, leadership didn't actually transfer ownership of the product's performance to the team. Leadership just put fear into them, and the teams responded as such. Fear suppressed the team's natural desire to create a reliable product that would make the customer ecstatic.

There's a type of leadership that often hands ownership to the technical team. I see this in startups. Teams in startups know that a product failure is likely the end of the company and all is lost. No one cares whose fault the failure is. There's no next program to get transferred to. It's just over.

But there is no reason large, longstanding companies can't have this as well. Simply change program dynamics so ownership is transferred, and there are several ways this can be done.

The following example shows how the goals were successfully transferred to the team. When the team owned the deliverable they overcame a tremendous hurdle and delivered exactly what the customer needed: new tech that was highly reliable and in the customer's hands quickly.

Successful Transference

I was invited to work with a team that was developing a next-generation product. It was centered around a laser system that did things I only believed existed in quantum physics books. I know lasers. I still couldn't believe this thing existed as I lay in bed that first evening. Manufacturing this system cost $2.5 million. Then, there'd be a substantial markup for anyone who would like one. They had someone who wanted one and they wanted it NOW!

The design wasn't entirely new. It was their flagship product with these recent advancements added. The new model worked great in the lab, but was it mature enough for the field?

The executive team knew me from several previous engagements. They wanted me to assist with this pressure-cooker of a project. Why was it so intense? Why did they want me to contribute at every level, from the technology side of things to its project management?

This project had the hallmarks of a project management nightmare. The program had to be completed rapidly, it couldn't be allowed to fail, and it featured new technology that was only marginally stable. That old adage "Fast, reliable, cutting edge: pick two" wasn't going to be upheld here. They wanted all three.

My assignment was to answer these two classic reliability questions: "When will the new features wear out?" and "What is the rate of random failure?"

The product was commissioned by their largest customer. Simply put, the stakes were high. Their product was used in their customer's production process. If their product failed in the customer's hands the production line would stop. And this production stop could be measured in millions of dollars per hour.

Obstacles to Transference

Here is where the insanity of this situation became evident. The effort to release this next-generation product – which held my client's very brand value – was in a program with an almost comical limitation on time and material for testing.

Simply put, I thought leadership was joking when they said we couldn't have a complete prototype system to test. No, they wanted their very important customer to receive the first unit ever produced. Otherwise, we wouldn't make the deadline.

For a split second I thought, "Wait, was I hired to run an impossible program and be a fall guy?"

This unrealistic program framework was created because the CEO didn't see the program the way I did. He saw a mature field-proven product that was getting "a simple upgrade." He believed that the calls for "all this demonstration testing" were just nervous engineers wanting to cover their butts.

The company had been producing products like this for decades and this one was 80% of a well-proven design. In addition, the new technology had already been shown to work on the bench in the lab, so what was the problem? Just put it in.

What the CEO knew was that a competitor was breathing down their necks, waiting to slip into their spot with the customer. He wasn't going to give them that chance.

So the product development team was left with no opportunity to explain that "just because a new technology works on a bench and constitutes a technology change of only 20%, this in no way guarantees reliability."

And here's the kicker. This product, which costs $2.5 million to produce, had 30% of its cost in the new technology components. You may not know this, but to demonstrate even the most minimal confidence in a reliability or life goal, you have to test not just one product to full failure but over 20! Clearly, this was not going to happen. We needed a plan B.

Successful Intervention

How do we do this? What do we ask for? Why am I sweating so much?

The test program wasn't going to work if the yeses and noes of reliability came from leadership. Whether the product was truly reliable had to come from the team.

If I hadn't intervened, leadership would have ordered the team to follow their edicts, and the team (having no choice) would have obliged. The result would have been an incomplete test program, yielding no genuine result. Reliability engineers call these "window dressing test programs." They look nice but don't accomplish anything.

It was clear to me that the only way to create authentic product confidence was by transferring ownership to the team. What, in this case, does "transfer" mean?

Remember, you can't transfer something you still hold. When it came to directing how to reach the goal, leadership would have to let go. The team would have to be given freedom. It shouldn't have to answer to anyone as to how it did what it did. The responsibility should be theirs.

In this case, the result was one of the most unique testing programs I've ever been a part of. The technical team knew where the risk was better than anyone, so they became the goal's new owners. The program they created was spot on. Following someone else's directive would have bypassed their valuable knowledge.

Objectives and Transference

I asked that the leaders and I have a 90-minute uninterrupted meeting. In addition, I asked that each leader have budgetary and schedule approval power. Simply put, we all needed to know that the decisions made in that meeting were immediately actionable. They obliged.

I began the meeting by telling the leaders that the ownership disconnect was costing them greatly in available resources. They assumed I meant they'd need to hire more people. That wasn't the case. They were simultaneously relieved and confused when I said, "No new hires are necessary." They were wasting resources by not transferring project ownership to the correct positions.

If they followed my recommendations, program results would match what a 50% resource addition would accomplish. They looked unconvinced.

I then said, "Have you ever observed how individuals seem to put out twice as much output in small startups?" We just accept that large organizations are sluggish. We tell ourselves it's an inevitable symptom of bureaucracy. It's not. It's an inevitable symptom of ownership floating up and away from the people who can take action.

True and Interpreted Objectives

The result in this case was a lack of connection between those with the most technical expertise and the true program objective, not the interpreted program objective. So what's the difference?

An interpreted program objective would be, "We're having people over. I want the lawn mown." A true program objective would be, "We're having people over. I want the front yard to look fantastic."

With the interpreted objective, the leader has extrapolated the required action from the actual desired outcome. Why are they the best person to do this?

If you've ever had a kid do yard work for you, you'll agree it can be an iterative process. I call it "the volley."

> "I'm done."
>
> "You have to pick up the sticks."
>
> "OK, am I done now?"
>
> "Don't just leave the leaf bag by the front of the house."
>
> "Can I go inside now?"
>
> "Put the rake away."

The Parent's Guide to Objectives

Because of what I have learned about leading programs, I have a bit of a unique parenting style. Although I'm sure it will be criticized in a therapy session someday, it is effective. This is an example of what I do different.

I give responsibilities, not chores. What does that sound like?

"You're going to clean the living room. Let's talk about what that would look like when you are done." I don't tell them to vacuum or remind them to move furniture or dust around window frames. I use terms describing the end effect.

"When you're done, the living room will feel comfortable to someone who is highly allergic to dust. You would be proud to invite your favorite pop star over and entertain them."

I also make sure they understand a volley is unacceptable. I will often finish with a statement like, "Failure is if after you are complete I have to provide input to get the result we agreed upon." I do allow passes if a short volley occurs due to a differing opinion on what finished looks like. But they have also learned that they are responsible for that too. They have learned to ask for specifics on what I think "complete" is in advance. They own the outcome.

This is an example of a group transitioning from interpreted objective leadership to true objective leadership. In fact, this is one of the most common I have seen within reliability programs.

Interpreted objective: "We need HALT testing completed." I get this all the time when I first engage with new customers.

True objective: "We need a design robust enough to have a 99.99% reliability. Even when used at 20% higher than the toughest use case."

It's that simple.

What Transferred Ownership Looks Like

How do things change after this? We see a diminishing trend in weekly status reviews, high-resolution Gantt charts, and the dreaded "daily standup meeting." All of those actions are a part of the volley. When they fade from a program it's a clear sign ownership has transferred. Who needs a volleyball court when no one is playing?

I'll share something personal. More than once I have had to hide a smile in a planning meeting where we are hashing through these strategies and I see it "clicking" for leadership. It reminds me of those dramatic moments in movies where the people in charge realize that the "regular guy" is going to be the one to save the day. It's a plot center point in so many movies.

- *Apollo 13*: the young design engineers are allowed to propose crazy configurations for a new oxygen system with existing parts in the capsule to save the crew.
- *Armageddon*: the blue collar oil drilling crew comes up with a better design than NASA for a drilling rig to save the planet from a comet.
- *Independence Day*: the fighter pilot figures out the aliens' weakness and executes a plan without permission.
- And of course, every gritty cop story where the detective ignores the chief's orders and goes with his "gut." It always ends with the police chief saying, "Damnit, Smith, you are a maverick and total pain in my ass, but one hell of a good cop!"

The moral of all those stories: good things can happen when ownership extends to individuals on the frontline. The writers know we have all been frustrated when leaders don't listen. That's why we love those stories.

The Benefits of Successful Transference

One thing was clear in this laser project with a budget for zero samples and a comically compressed schedule. Only the technical team knew enough to identify the risks with surgical precision. This precision was a necessity if any reliability was to be demonstrated in a program that was so constrained.

With the ownership now transferred to the team, they were free to apply an approach they knew worked. They specifically parsed out what was driving the highest risks. What was now absent was the fear of not doing it right. How they got to great results was up to them. All that would be measured in the end would be the outcome.

Do you know what I found to be strange? The dynamic of results ownership by the engineers and scientists had already existed. In fact, it was just in play the previous week. Leadership had totally trusted the team to invent the technology; they never told them how to do it. Why when they switched from invention to reliability did the team go from driving actions to receiving instructions? I suspect this is driven by a sense of accountability of great reliability performance by the leaders. It's true they have this accountability, but why take that accountability away from the individuals who can do the most about it? That's the strange behavior we see in programs again and again.

What happened next was amazing. The team developed a sub-assembly that did not represent any part of the product. It's only purpose was to demonstrate the highest risk to poor reliability. When the test systems didn't have to be similar to the product, the team was able to optimize for cost and speed while still demonstrating the new technology's reliability.

A Racing Bike Analogy

Turns out we didn't need that actual laser system to test if the new features worked. We could replace the laser response with other metrics. How did we get there? One simple reason. We trusted the individuals who knew the physics.

If it isn't clear as to why this worked so well, here is a good analogy. Suppose we were manufacturing racing bikes. We've finally cracked the code to building nano carbon fiber wheel spokes, ultralight, like nothing the bike racing community has seen before. We next need to test our new invention.

First question. Racing bikes are very expensive, and we need over a 100 to do adequate testing. Does testing require an entire bike? No.

If the team has ownership of the deliverables you might find a solution like this. Engineers riding a child's toy wagon with bicycle rims on it down a poorly paved hill. It looks stupid, but with that setup and a set of both new and traditional style spoked rims the reliability comparison is fast, easy, and cheap.

The CEO of Nano-Carbon Spokes LLC is going to need big trust in his team. He will be deciding to go forward with production based on results from a test code named "The Rusty Wagon on a Hill test." But I bet it would be a good decision.

The laser team came up with something amazing: their wagon. They created a small assembly that cost $2500 to manufacture. That was not just the cost savings on the scale of a wagon vs a racing bike. That was a used wagon bought at a yard sale late one Sunday afternoon when the owner just wants to get the yard cleaned up vs a car. I was in shock!

These units could be manufactured locally and delivered in 15 days. The days of strategizing desperately how to get just five test units turned into "Yeah sure we can make 200. Is three weeks, OK?"

Guided by All the Goals All the Time

Many leadership books talk about the critical need for roadmaps. Going all in on roadmaps, however, can be dangerous. They can cause tunnel vision. You overfocus on the map without paying enough attention to other things. You fixate.

The Roadmap Conundrum

With a roadmap, what you need is a perspective that's balanced. Focus on one part of "the route" too long and you're going to throw things out of balance and create a mess.

Imagine a car trip. To help you reach your destination, there are many critical elements: the windshield, gas gauge, speedometer, directions, and so forth.

What's most important is the balance you give these different elements.

Each element has its time and place for use. Take the windshield: 99.8% of the time, you should be looking through the windshield. But, say, you want to change lanes: you need to take your eyes away from the windshield for a second, so you can look through the rear view mirror.

The balance between elements needs to be respected. You can't decide to give one element more attention than it deserves, because if you do you'll get lost, run out of gas, or even crash.

Obviously, you and I would never operate this way in driving a car. But for some reason, when the responses are far removed from the inputs, we can forget our common sense and run a program that way.

Overfocusing on a program plan when setting out to accomplish a goal-based program has a similar type of effect. It's a slow-motion crash with injuries that just take longer to become evident. So it's possible to use this ineffective method for a long time.

Why We Embrace Tunnel Vision

It's possible to never make the connection between our myopic perspective and the outcome. Humans are amazing at doing this. We probably developed this skill to forget pain, just to make it through a day when we lived in the forest, were dying from starvation, had an infection from a stick we stepped on, and found a grizzly had taken over our cave. Life sucks and sometimes a myopic perspective is what saves you (that was before beer was invented).

When No One Has a Plan

We have also seen programs that suffer from the opposite condition: a weak plan. This is the equivalent to driving without directions. Using stop lights, road markers, and dash gauges as the sole navigation tools is a crappy plan if you need to be somewhere. Where that trip ends is anyone's guess, but it is very unlikely to be where they hoped to be.

The title of this section is "Guided by All the Goals All the Time." The word "All," used twice, is what makes it a meaningful statement.

Ignoring even one of those inputs from outside, inside, or the plan, during the trip results in disaster.

The complete set of information required to guide the program should be streaming in synchronicity all the time. We use this information stream to make complex decisions continually. If the process decays and starts reacting to singular pieces of information, we'll wildly swerve out of control.

This state is evident because "emergencies" keep occurring. I imagine this to be like waiting until we hear the bumper scraping the guard rail before steering back towards the center. In addition to the unfortunate damage, we jerk the wheel. This sends us careening toward the opposing guard rail. We jerk the wheel again, and repeat.

Kneejerk Reactions

That's how programs are managed when we aren't monitoring all the goals all the time. Guiding by all the goals all the time is a driver who just smoothly maintains the center of their lane. It almost looks effortless, like they aren't doing anything. But in fact they're making thousands of small corrections continually.

An example of a program with the elements not correctly monitored is when we have a spike in field failures due to a rushed design process. Someone may have given time to market too much attention earlier. What's the kneejerk reaction to this? Pull resources off of the new program to support fixing the field failures. What's the consequence?

The new program is now off track, under-resourced, and behind schedule. If leadership doesn't let up on the time to market, we can expect a poorly developed new product to be sent to the customer.

Are we really going to act surprised when the new, underdeveloped, product has a sudden spike in field failures? Of course we are. Wash, rinse, repeat.

Summary

The problem here is not just about being frantic. The lack of efficiency is brutal, and grows exponentially. Getting programs back on track is 20 times more resource-intensive than if the same tasks were planned.

For some reason we decide this is all normal and just keep moving forward. It was all in such slow motion that we didn't connect the poor profits, top talent leaving the company, and lost market opportunities as the casualties of the poorly steered process.

The solution is that all the goals have input through the entire program. There are two key elements to this. The first is the entire team absorbing responsibility for the complete program goal. The second element is the Bounding methodology that ensures all of the needed inputs for even the smallest steering input are present all the time.

We talk about the Bounding method in upcoming chapters. You're really going to like it. It's easy to apply, and you can put it in motion immediately.

References

1. Drucker, P. (2011). *The Practice of Management*. New York: Routledge.

6

New Roles

For an organization to deliver reliability effectively, it must create a few key roles. Most of these roles will be temporary, and will only exist for the life of the program.

Which roles am I talking about? These three:

- **Change agents:** they're responsible for changing how the organization operates.
- **Reliability czar:** they're responsible for communicating between top-level leadership and the frontline of product development.
- **Facilitators:** they're responsible for how a particular initiative gets executed.

None of these three roles is its own job. In other words, you're not hiring change agents, reliability czars, and facilitators. Each role is filled by a current team member who already has another primary role, like engineer, department manager, or program manager. They're ready to put on one of these secondary hats, however, when the need arises.

Let's look at each of these roles, one at a time.

Role of Change Agents

The change agent is a team member who brings a fresh perspective from one organizational level to another. This could be the perspective from, say, an executive to a technical lead or from an engineer to a director.

Whether it's carried out by a single individual or by many, the change agent's role shouldn't be underestimated. While they can't necessarily forewarn of a disaster (that would be the role of an oracle), their role is to bring back the right information at the right time. Information that assists with decision making.

A change agent never demands a deliverable. Instead, they make change happen by providing information and shaping the playing field.

Change is hard, personally and professionally. It's hard because we accomplish 70% of our tasks while on autopilot. That of course makes sense. If we tried putting deliberate thought into our daily tasks, we'd quickly be in tears from overwhelm.

This autopilot dynamic becomes clear when we look at how it relates to driving a car.

Reliability Culture: How Leaders Build Organizations that Create Reliable Products, First Edition.
Adam P. Bahret
© 2021 John Wiley & Sons Ltd. Published 2021 by John Wiley & Sons Ltd.

Take the student driver. Behind the wheel, they look like fawns in a windstorm, standing on ice, because they haven't built their autopilot abilities yet.

Then there's the experienced driver. They've driven so much that their autopilot capabilities have reduced conscious thought down to almost nothing. I myself have driven to a destination with no idea of how I arrived there. Absolutely zero. I imagine you've done something similar. We've driven so often that we can do it 99% on autopilot.

At work, so much is automatic, too. What we need to be aware of, then, is this: if the challenge of change isn't taken as its own initiative, the change effort will fail.

It's human nature to continue on our default behavior paths. That's why there's nothing easier to sell in this world than a quick way to lose weight or build muscle. We all know that the only way to accomplish those goals is through diet and exercise. Yet, the habits of eating and exercising are ingrained deeply in our psyche. The change process here takes some serious uprooting.

After diet and exercise, workplace habits come in third for "difficulty of change." That's because we're at work half our waking hours. There's no way we could handle that if we didn't have an autopilot at the controls.

To change our work habits, we need the equivalent of a personal trainer. That's the job of the change agent.

For this to work, we do in fact need three different "personal trainers," or change agents.

The first is the executive change agent. The executive agent's goal is to make sure that all the leaders involved in the project take complete and personal ownership for the product's reliability. After all, if the top leaders don't champion reliability, how can it stick?

The second agent is the project manager (PM). A few years back, I was working on a project and asked the PM, "What are the program's three most important goals?" Without hesitation, she said, "The first goal is 'schedule.' The second is 'schedule.' And the third? 'Schedule.'" She was making a joke, or maybe not – I'll never know. But that joke revealed a lot about the PM's psyche. That is, PMs are willing to sacrifice all the other program goals if it means that the schedule will be upheld. It is essential, then, that you get the PM on the side of reliability.

The third agent is the reliability manager. If they want reliability to take its rightful place in the organization, they can't just focus on reliability alone. They can't be selfish. They must show that they support the other disciplines and their goals, even as they work hard on keeping reliability front and center.

This is a sort of holy trinity of support for change. Now, and at all levels, separate areas of an organization that would normally come into conflict with each other when change is initiated will in fact have to coordinate to make the change happen. Change itself has become the goal.

Reliability Czar

The czar is cool. They're a conduit through, or even a portal to, the organization chart. They make sure critical information gets shared between departments and levels, so important packets get to the places they need to be, immediately. A czar, for instance, will take a message from lab technicians and make sure it gets directly to the CEO. Without the czar, such a seamless transfer of information isn't possible. Especially when it comes to jumping levels like that.

An important part of the czar's role: they need to be steeped in technology. Why? They'll need to have meaningful conversations with engineers, technicians, scientists, and designers, and unless they're technologically sophisticated themselves, these people won't want to speak with them.

The czar also has to be trusted by the organization's leaders. I should be clear in who I mean by "leaders." The leaders I mean are decision makers. They don't have to check with anyone to make decisions regarding schedule or resource changes. If a leader asks that two months be added to the schedule or $100 000 be added to the budget, that's what happens. A leader may be the CEO or the president. They could also be a VP of R&D. It all depends on how the organization is set up.

Let's look at the czar's role more closely. It serves three functions:

- **Function #1:** it acts as a link between the test bench and leadership, so observations, discoveries, and issues can be easily shared.
- **Function #2:** it helps leaders give the hands-on team members direct input.
- **Function #3:** it ensures that the raw information is distilled down to its essential elements. (How does this happen? Well, if information is being passed back and forth, the parties involved have to get good at finding and articulating the crux of an idea, so it gets shared accurately.)

I think it'll pay off if we examine each of those functions, one at a time.

The Czar is a Link

Let's start with that first function: "acts as a link between the test bench and leadership." Why is this function necessary?

As reliability information travels up the organizational chain of command to the VP of Quality, the VP of R&D, the VP of Service, and the VP of Manufacturing, it's natural that some of it gets lost. That's because midlevel managers naturally filter the information – sometimes unintentionally, sometimes intentionally. As it gets passed upstairs to the executive team, it changes. (Think of it as being like the game of telephone.)

The czar combats these kinds of changes. They realize that sharing distorted information is inhibiting. Sure, the genuine unaltered information is full of grit and potential conflict, but that's what makes it valuable to a leader. The unaltered information is what helps the executive team make critical decisions.

The reliability czar is not a glamorous job. It's about making sure that important information gets handed to leadership. It's not about making decisions.

When you're a czar, you're going to get dirty, and no one is going to thank you. Some might even try to stop you. But you may just prevent a disaster, once the product hits the market.

It's important to note: the czar isn't a new executive position. It's not "the VP of Reliability." We don't need that. The czar is the responsibility of someone lower down the organizational chain, who has been deputized. Once they put on their czar hat, though, they don't have to check with their boss as they pass information upwards.

Direct Input

OK, onto the second function: "helps leaders give the hands-on team members direct input." What's this about?

It's Function #1, but in the opposite direction. Function #1 was about how the czar and project team gets information to leadership, and Function #2 is about how leadership asks for and gets the information they want. It's push vs. pull.

The second function of allowing the decision makers to have direct input to the hands-on team is critical. That's the reason CEOs and presidents should be particularly interested in creating this position.

Recall that story about the VP in Chapter 4? That VP couldn't find out why the engineers and technicians wouldn't test the new sub-systems, no matter how many times he asked or how serious his threats were. That's a frustrating position to be in. Even after the project, he still couldn't figure out why he was always in the dark.

If he had had a czar, he would have found his answer within a few hours of having asked the question.

In many organizations, a request that a leader makes will be translated two or more times before being received by the individual who can take the necessary action. That's a lot of filtering and adjusting for personal interests before anything meaningful happens.

Distilling Information

The third function is about collecting and distilling information. Some people call it "data scrubbing." You don't change the information, but you organize it and make the discrete messages clear, so that leaders can understand them. This, too, is the czar's job. They don't pass along an informational blob. They present it in a way that makes sense to the leader. Figure 6.1 shows how the czar transmits information.

Who is the Czar?

So how do you pick the czar?

The czar shouldn't be a midlevel manager. They shouldn't have another leadership role. Instead, they should be an individual contributor or, at most, a first-level manager who is an active technology contributor.

The czar should speak the language of "The People" and then learn the language of the "Leaders." Doping it the other way around is too difficult, because the language of "The People" is more complex and the leaders know a bit of "People" language and can close the gap on any poor translations. Many of those leaders were once commoners themselves (no offense).

One of the reasons I personally bailed on the fast track to the top and became a consultant is because I never wanted to be isolated from what was really happening in product development. I love it too much, and only hearing about what was happening out there on the frontline in weekly summaries just sounded like misery to someone like me.

(To stay sane, I always have three major technical projects going on at home. How many more would I have going if I got none of that stimulation from my work? It would be ugly.

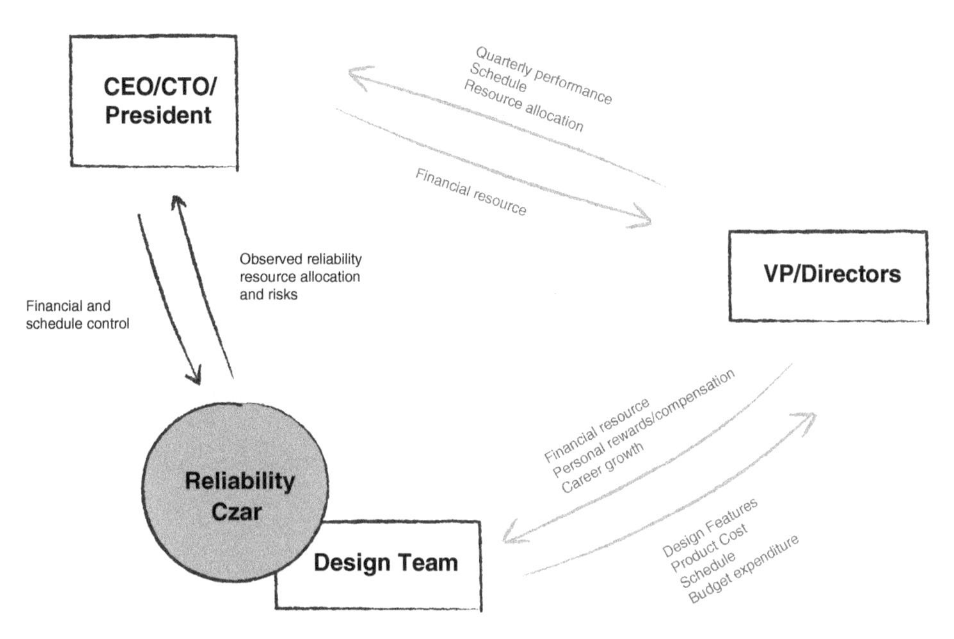

Figure 6.1 Reliability czar.

Divorced, living in a garage with four active car projects, one disassembled airplane, and in pajamas which I wear for days on end, which are in fact just my old racing suit. Like I said, ugly.)

OK, back to the czar. Their unique value is that, as someone on the frontline each and every day, they have the ear of the top decision makers. With this short cut between the frontline and the top floor of the command tower, we have a key piece to the reliability puzzle. The executive leaders get a direct path to have their concerns about long-term reliability shared with the people who can immediately do something about it.

I am not advocating just eliminating the information flow between the midlevel program managers and the executives. What I am doing is openly acknowledging that the goals and constraints put on middle management is not the best path for *all* reliability information. Those roles are highly responsible for program completion date, budget, and quarterly profits. Any good system has opposing forces creating balance. These forces have to have the same strength for the system to work. That is what we are adjusting for reliability with the czar.

This reliability czar will have a primary role in the organization and the project. It may be the lead mechanical, electrical, or software engineer on the project. They could have only five years of experience or be a principal quality person. The right person may be a 30-year manufacturing engineering veteran. This individual should actively engage with the engineering process in their primary role to some degree. This reliability engineer/manager may still be responsible for all the short-term goals of developing new features, staying on schedule, and hitting the cost point. They should be accessible for information from the leaders regarding the program for a year or two after product release. They may be on a new project but the connection to the previous product should not be discarded by the czar the same way it is not discarded by the VP of R&D or VP of Quality. That continuity for long-term performance is a gap we have to close for reliability. This is where the return on investment (ROI) pays out, and if it is not linked, it becomes increasingly difficult to justify investment.

It may sound like the czar is just left with this additional accountability dropped unfairly on them while the rest of the team runs off to new ventures when the program completes. But in fact they have been given one of the most powerful tools in the product development process. That commitment to long-term accountability and the ear of the decision makers makes them a voice that demands respect and inclusion.

Team members don't want to be the person that neglected a well-supported reliability initiative that has long-term benefits to the individual who is counsel to the CEO or president. It's a powerful portal. The reliability czar carries an extra load for the project but in turn has a superpower that no one else has. A power that not their manager, or their manager's manager, has.

How the Czar Works with the Team and Leadership

The team, the frontline, must believe the information being shared is providing valuable input that will assist with the goals that they worry about when they are driving, try to fall asleep, and of course obsess on in the shower until the hot water tank going empty jolts them back to reality.

If the team doesn't trust that the alliance of the czar is with the team then the role will quickly become detrimental to the reliability initiative. What about leadership? the role can also be ruined on that end if not handled correctly. Leadership must not feel the role of czar and the direct reporting it relies on brings conflict to relationships with, and between, mid- and high-level leaders. If the leadership begins to find that the czar's role is creating a level of distrust or conflict with VPs and directors, the effectiveness will quickly degrade. If this occurs, the demise of the role won't just be the dismissiveness of the CEO. The mid- and high-level leaders will begin to limit the czar's access to information where necessary information exists. You can't sample the water at the bottom of the hole if someone throws a big pile of gravel in there.

If the czar does not appear effective from the technical team's perspective, the soldiers on the frontline, then the czar risks transforming into a "mole" in their eyes. The following three steps can be implemented to remedy this situation:

1) Allow the team to do the final edits on any distilled summary information that will be shared with leadership. This option can sometimes slow communication but that is a worthwhile tradeoff if there is a concern about trusting what is being reported.
2) Allow a change in team member who is in the czar role. By doing this the element of "personal gain from access to leadership" is kept in check. It can be perceived that the czar may be using this special privilege for personal gain by appeasing those who have power. This could be in the form of telling leadership what they want to hear. It could also be simply sharing information that the team is unaware is going beyond the immediate group.

3) Permit anonymous votes of confidence in the czar's ability to transfer valuable information to leadership. If there is a vote of more than 40% "no confidence" then the individual in that role should be changed.

There is also the scenario where leadership trusts the czar but feels they are not providing valuable information. Like when my daughter used to interrupt me when I was on an important conference call to tell me things like, "Some cats like milk but Nicole just said her new kitten won't drink milk."

I am sure that is a factually correct piece of information but talking with the VP on the other end of this call about why the FDA wants to know why our "essential performance factors" don't match the definition in Standard 60601 is the conversation I need to be having right now. . . I just say, "Hold on. That's really unusual, honey." Steve's either not going to take my counsel too seriously going forward or wonder why I called him "honey."

Leadership has skin in the game with having the czar because they are potentially creating some strain with their direct team. The czar is someone providing information that their team cannot easily control or filter. The return must exceed the cost to be maintained.

For leadership I can't give any specific guidance on what to do. It is simply a personal call if you want this role to continue in your operation. It may have greater value in some programs than others. With regard to the value of the czar to leaders it is up to you to monitor and adjust if they perceive the role as valuable. One metric to monitor is the push vs. the pull of information. It is a valuable relationship if leadership is "pulling" information. If the relationship transitions to the czar "pushing" information on leadership then it needs to be re-evaluated and adjusted before it is terminated.

I discussed this dynamic a bit in an earlier section but it is worth reviewing again. The difference between a leader "pulling" information and having it "pushed" is simply this: if leadership is "pulling" information, a majority of what the czar is sharing is based on requests, from leadership. If the czar is providing standardized reports to leadership for most sessions then they are "pushing" information. I was once told by a mentor that unsolicited advice is of little value. It is only serving the person giving it. Advice you have solicited is what you want and so has high value. This is very relevant to this relationship because it is a relationship based on advisement.

Tips for the Czar

Remember that distilled information that is passed comes from many sources.

There is great information in all those little comments that float around day to day. Find a way to get all those little complaints that happen person to person to coagulate into a voice that can be heard. The czar is not just sharing their observations: they are the voice for their people.

Develop projections. Have the team project the impact of decisions regarding reliability resources in advance of big program redirections. Information includes interpretations of the data. This is where the value is at: smart informed people providing input. This information on demand for an interested leader is very valuable to a program making quick corrections to a changing landscape.

Make it easy for team members to participate. These individuals are so overflowing with passion to express their frustration that you will have no problem getting them to take a few moments to fill out an anonymous survey on the impact of reliability tools to specific issues.

To make polls work, minimize the amount of permission needed. Make it a game – keep it informal. There is some caution to be had when commentary is done on past projects where leaders made decisions and bad results occurred, i.e. blame.

Change isn't easy and especially in the game of reliability where we measure results in percentage of failure. We rarely measure reliability in percentage of success. Let's be honest, the best we can do, truly, is very small failures. So tread lightly: these methods can show benefit to the leaders who can make the changes happen if it is received well.

Role of Facilitators

Facilitators are often undervalued. A facilitator should be the captain, but are often demoted to scribe. As the team's captain they conduct how the team interacts with the analysis tools and each other. What's a ship without a captain? A group of people doing stuff aimlessly. That's what we see happen when an initiative doesn't have a facilitator: nothing. An analysis guided by committee is rarely a good thing.

One of the reasons we create teams is that they provide differing opinions. This is a beautiful thing. Differing opinions create diverse information. They create more information, which is good. But we have to do something good with all that information, but left untethered these differences are a chaotic collaboration with tailspin after tailspin.

The facilitator is fundamental to ensuring the varying perspectives of the team are woven into a single directive for the program. A team with a facilitator has a captain who coordinates and extracts the valuable difference of perspective and insightful input.

A facilitator may drive a workshop, a Failure Mode Effects Analysis (FMEA), a brainstorming session, or a root cause analysis session. Each of these activities has specific deliverables that are outlined in advance. The facilitator's objective is to have arrived at those deliverables at the session's conclusion.

Facilitation Technique

When facilitating, there are a few important techniques.

During a Design Failure Mode Effects Analysis (DFMEA), where should the facilitator place their focus? They need to get collaboration from a group of analytical individuals. That's really hard. Really, really hard. Each person is deep in their own head and following an idiosyncratic train of thought. The facilitator has to carefully extract that thought, so it can be examined without unraveling.

Pulling those trains of thought out of each person's head and dropping it on the table in a manner others can understand is no small task. In the DFMEA, there are elements of data and technical analysis, vested interest in outcome, and pride. Without a leader navigating the group toward a common goal, an effective outcome would be impossible. It may not be evident on the surface, but this is creative brainstorming.

The facilitator has to read personalities. They can't let one person dominate. Especially an extrovert. If an extrovert is untethered and left unchecked, the whole session can go down the tubes. It's not that the extrovert is being a bully. It's just that they can seem overpowering by their sheer natural enthusiasm. The extrovert expects others to respond as they would. Just interrupt and blurt it out.

A talented facilitator will pull the extrovert back in a way that doesn't shut them down. The facilitator will interlace contributions from extroverts with participants who are more reserved and need coaxing.

We don't want to shut the extrovert down. What a good facilitator aims to do is get the other more reserved members interacting in a manner similar to the extrovert. It will just take a combination of the facilitator's aggressiveness and the introvert's insight.

An introvert may give quiet visual cues that they want to participate, or are frustrated with the dialogue. By nature, introverts are usually paying close attention to body language and nonverbal clues, and expect others to do the same. If an introvert strongly disagrees, they may look away or roll their eyes. This may not be effective and may go totally unnoticed by extroverts.

A good facilitator scans the room and observes body language constantly. They're not looking for participant happiness. What they're looking for is a change.

A change in a participant's body language could indicate a new thought or emotion, and it's the facilitator's job to decide if that change should become part of the discussion.

Some people sit with their arms crossed and a stern look on their face even if they like what they're hearing. Others will smile and nod even if they think you're speaking garbage. Some signals are subtle, such as an individual's blinking multiple times when they're uncomfortable.

There are a few common introvert and extrovert attributes:

Introvert Checklist:
- Works best when alone.
- Listens in groups and does not regularly share their own ideas.
- Energy is drained by group interactions.
- Speaks concisely.
- Needs advance notice for requests or assignments.

Extrovert Checklist:
- Includes feelings and background thoughts when communicating.
- Speaks before thinking or works out idea while speaking.
- Looks to others for inspiration and enjoys group work.
- Energy level increases with socialization.
- Willing to work on a task "live" with no advanced notice.

To be a useful facilitator, you may need to be a bit different than you are normally. I don't mean being a phony. I just mean that you might have to be a bit more directive, like a traffic cop. If making that kind of change strikes you as abrupt, you could warn the attendees ahead of time. I often start a facilitation session by letting people know that the nice friendly Adam is going away for a while. It's "facilitating Adam" who is going to lead the session.

In my role as facilitator, I'll interrupt and instruct others to contribute when they don't want to. I may refer to earlier parts of the conversation and insist we discuss them. To an outsider, I might look like an alpha jerk. That's OK, though. You have to put aside gentle ways to get the job done.

Creating a Narrative

A facilitator also translates all the dialogue into a coherent narrative. In essence, they're functioning as a live editor. That's one of the most important reasons why the facilitator is *not* the session's scribe. This may seem counterintuitive. But it's disabling being the scribe. The facilitator is too busy keeping track of storylines and making sense of the whole to get bogged down in minutiae. A separate team-member-as-scribe is mandatory.

In sessions like DFMEA and brainstorming meetings, the facilitator is actively trying to extract unique input. They are also analyzing the input in real time, looking for patterns. The input should at times conflict. Universal agreement among team members is suspicious. Sometimes this conflict creates even more input – this is good! The facilitator may even stir up conflict, but then distil it to what's essential and most valuable.

The facilitator has to sense when the play isn't advancing. If there's a diminishing return on the energy and time going into the debate, it should be stopped and taken offline. The team's momentum can be slowed if 80% of the team is being dragged through a discussion they don't understand. The interested parties can continue their discussion offline and bring the conclusion to the next session.

The facilitator is also the PM. They have to manage the initiative's most critical resource: time! If time runs out, the team won't be able to complete what they need to. Time is also a cruel mistress. More time lets you do more. More time also drains morale. We all have other things to do.

It's astonishing how much energy it takes just to be in a meeting. Simply sitting in a room and thinking can be exhausting. I guess four walls and humming fluorescent lights aren't natural. So read the room and suggest breaks when a recharge is needed.

A difficult part of the facilitator's role is getting the group started. All good facilitators have a set of starter questions to trigger engagement. These questions can also help you keep the participation balance between extroverts and introverts. Engaging the introverts with intriguing questions should have them firing off ideas at the same rate as the extroverts.

What are some common starter questions? They fall into different conceptual groups.

You have observational questions: "What about this issue do you notice?" and "What deliverables are we aiming for?"

You have reflective questions that push individuals to think instead of simply observe: "What does this remind you of?" and "What about this issue do you find problematic?"

Then you have interpretive questions, which also get participants to evaluate the task at hand: "What does this mean?", "What could be different?", and "What more do we need to know?"

Role of Reliability Professionals

We just spoke about three secondary roles within an organization. It's worth also taking a moment to discuss how the base role of the reliability professional needs to change.

Much of this book is aimed at asking leadership to change how programs are run. The rationale here: if the program's reliability portion is run poorly, it's going to cost the company big money in the long run.

But it's the reliability managers and engineers – who we'll call "reliability professionals" – that are a key element in this change.

Stop Asking for Resources

The reliability representatives have to stop asking for resources blindly.

This book was written for executives who want an organization that makes highly reliable products. The executives need to also understand what a reliability professional looks like, from the inside.

A fundamental problem with a reliability professional simply requesting resources is that from their spot in the organization they can't see the big business picture. Who can see the big picture? When the reliability professional is the reliability change agent they are then privy to the information to provide requests that consider all facets of the program.

The reliability representative change agent is at the ideal vantage point to see the impact of both doing and not doing a reliability activity. Transforming from a reliability nuisance to a reliability change agent requires one simple shift. Provide information, not requests.

This is in contrast to what is often prescribed to group leaders. "Don't bring problems, bring solutions." But to suggest a solution when you are not at a vantage point to see the full problem doesn't always help. What needs to be identified is what information will assist the leadership to make the best decision. This is often in the form of a quantitative assessment of risk with making a decision.

Connect Reliability to the Market

Historically, many of the reliability missteps occur because risk was not quantitatively assessed. It was simply a cumulative gut feeling from the midlevel leaders in the room. The problem with gut feelings is that they are subject to change week to week, day to day, and meeting to meeting.

A quantitative risk summary has dollar signs on it. Dollars is the only metric that can be used at the highest program level. There is no other way to correlate schedule goals, R&D budget, quality goals, and program cost target. Dollars are the universal adapter for all of these to relate.

Connecting reliability to market share is difficult because it has an intangible element: feelings. If a customer expects high reliability and gets low or medium reliability, how does this affect market share? Well, if they are disappointed it is less market share; if the reliability exceeds their expectations, this should contribute to more market share. OK, good: we agree on these two truths. But how do we make it quantitative?

There are a few factors to estimate:

- Market sensitivity to reliability.
- Customer margin on reliability expectations.
- Reliability impact on product's usefulness.
- Customer dependency on the product.

The market sensitivity to reliability can be best characterized as to how likely a reliability issue would rank #1 in categories of customer dissatisfaction. Let's do an example.

The product is clothing. Three categories for customer dissatisfaction could be fit, style, and durability (reliability).

This is the question we are asking. If the customer were to experience a slight issue in all three of these categories, what order would they list them in a poor review?

Here is that review:

> I purchased the dress shirt and found that in the shoulders it was too tight. The style didn't look as nice as it did on the model in the advertisement. After wearing/washing the shirt four times I saw that some of the thread in the sleeve was loose. I do wear the shirt occasionally but I would not purchase from this seller again.

What about a phone? What is most important? What is listed first?

> The screen goes blank at random once every month. If I give it a full charge again it comes back. I found that it processes slower than some of the other phones I have owned. The outer coating is peeling after just four months of use. Total crap, can't use it, don't buy. I replaced it with another brand.

Reliability is everything with the phone. Even if the reliability issues are intermittent and don't require service, they are a deal breaker. I can't be on an important phone call and have it drop. If I use an app to navigate and the phone dies, I'm screwed. I usually don't even know the address I was headed to.

Comparing these two products (Figure 6.2), it's clear to see what is more important. They are different and should be comparatively scored if we plan to use them to guide program decisions. By correlating a scoring system to program guidance we can avoid getting it wrong.

Shirt Customer issues (Greatest to Least)
1. Shoulders too tight
2. Style was underwhelming
3. Threads came loose

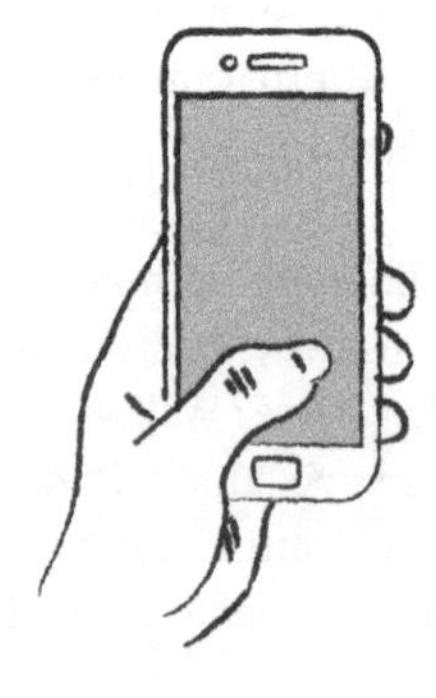

Phone Customer issues (Greatest to Least)
1. Screen goes blank once a month
2. Slow processing
3. Outer coating peeling

Figure 6.2 Shirt and phone ranked factors.

Summary

Roles are critical for any initiative with a team and a goal. We all know that any personal or professional project can come to a complete standstill if we are both stepping all over each other and at the same time there are important actions being forgotten.

Product programs, of course, have roles. What I am asking in this chapter is do you have the right roles? I discussed the importance of three specific types of roles. These don't have to be applied. What is more critical is simply evaluating the process and seeing if the tasks needing to be accomplished and ownership required match what people are being held responsible for.

7

Program Assessment

Measurements

"If you want something to improve, measure it."

I first heard that saying when I was a new engineer. I'm not certain of its origin, but it's clearly a derivative of Peter Drucker's phrase, "What gets measured gets improved" [1].

To a new engineer, the idea of simply measuring something to improve it sounds a bit simplistic. Similar to the homily, "An apple a day keeps the doctor away." That's because young engineers tend to think in logical – not emotional – terms.

To my surprise I soon found out that it was in fact true. Apples are magic! No, no, no. I mean that measuring does indeed create improvement.

Well, actually, both statements – the one about measuring and the other about apples – are true and they're true for the same reason.

They get us to focus, and once we focus on something we can make that thing better. It sets off a chain reaction of steps that brings about improvement.

If you eat an apple a day for reasons of good health, it forces you to think about your health for a moment each day. Doing so will almost certainly improve your health. "If I'm bothering to eat an apple instead of a candy bar, maybe I should go for a walk to get the most out of the apple?"

It's the same thing in engineering. The way to improve things happens as a byproduct of human nature.

If a robot is tasked with measuring something, it simply provides the measurement. If a human is asked to measure something, that person will provide the measurement and also be curious about what the value should be.

Once they find out there is an optimum value of what was measured, they will likely want to achieve that value through improved performance.

Measurement was the spark, and human desire for excellence is the fuel.

What are some reliability performance types and how would we measure them?

Reliability Culture: How Leaders Build Organizations that Create Reliable Products, First Edition.
Adam P. Bahret
© 2021 John Wiley & Sons Ltd. Published 2021 by John Wiley & Sons Ltd.

What to Measure

If what is keeping you up at night is your entire product line wearing out and failing in the field, then you should implement an Accelerated Life Test (ALT) (Figure 7.1).

If the customer is being too rough or the operating environment too unpredictable then measuring with a Stress Margin Test (SMT) will provide the information you need. "Will the product still perform in these extreme conditions?" Figure 7.2 shows how the lack of margin for a specific assembly will drive a percentage of failed population. The portion of the population with the weakest features used in the highest stress application will fail.

Often with SMT results the areas of risk are addressed in two ways. The first is by creating more margin. This can be done by moving the distributions further apart. Moving the distributions further apart can be done by increasing product strength or reducing the range of applied stress.

The second is to tighten up one or both of the distributions. This would be measured as a lower statistical standard deviation in the distribution. Reducing the standard deviation can be done by improving manufacturing quality or reducing the variability of stresses that can be applied when the product is in use. Figure 7.3 demonstrates these two strategies.

Using Reliability Testing as Program Guidance

In your product development program, how do you use reliability as a guide? You need a handful of reliability measures that you can monitor on a dashboard. This way, you'll know what to pay attention to and you'll know where you're going.

My favorite reliability measures? Try these two:

- The primary wear-out failure modes.
- The random failure rate during life use.

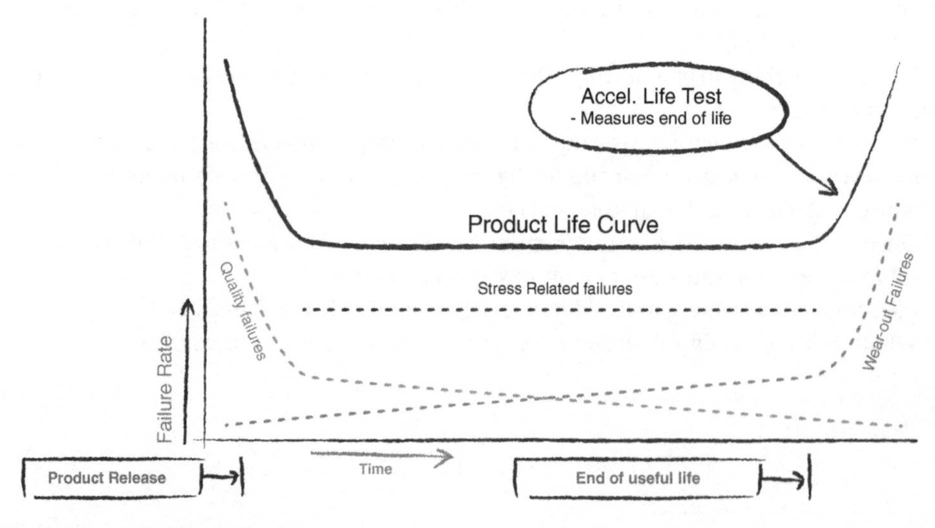

Figure 7.1 Reliability bathtub curve.

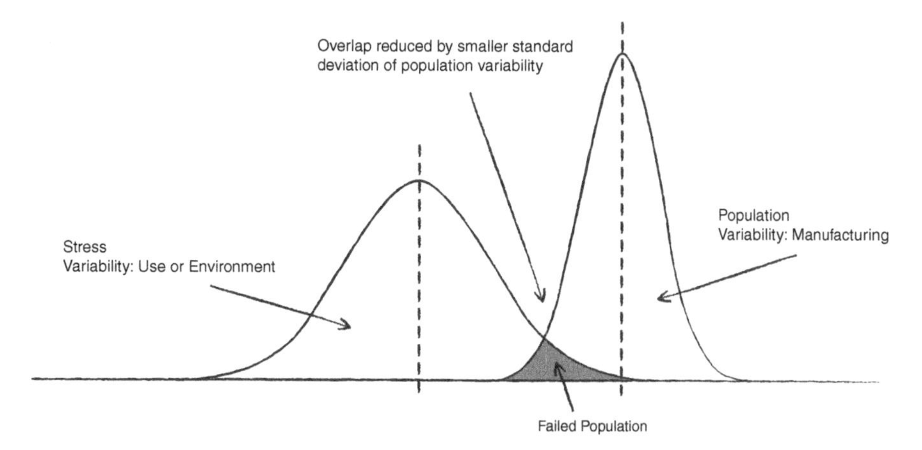

Figure 7.2 Stress strain overlap.

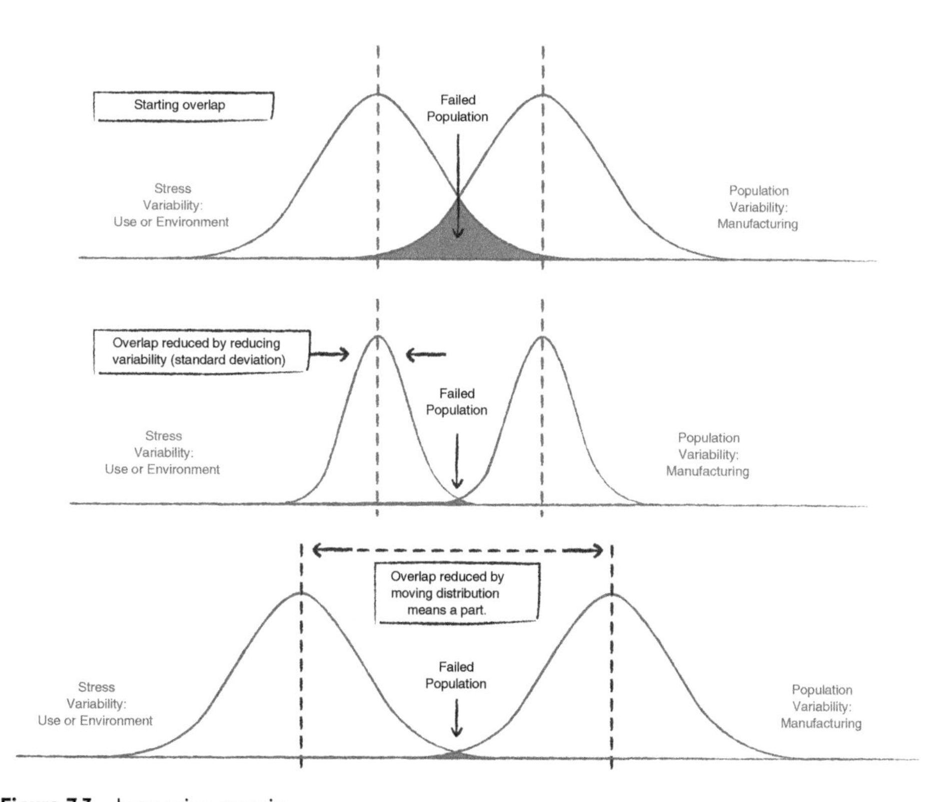

Figure 7.3 Increasing margin.

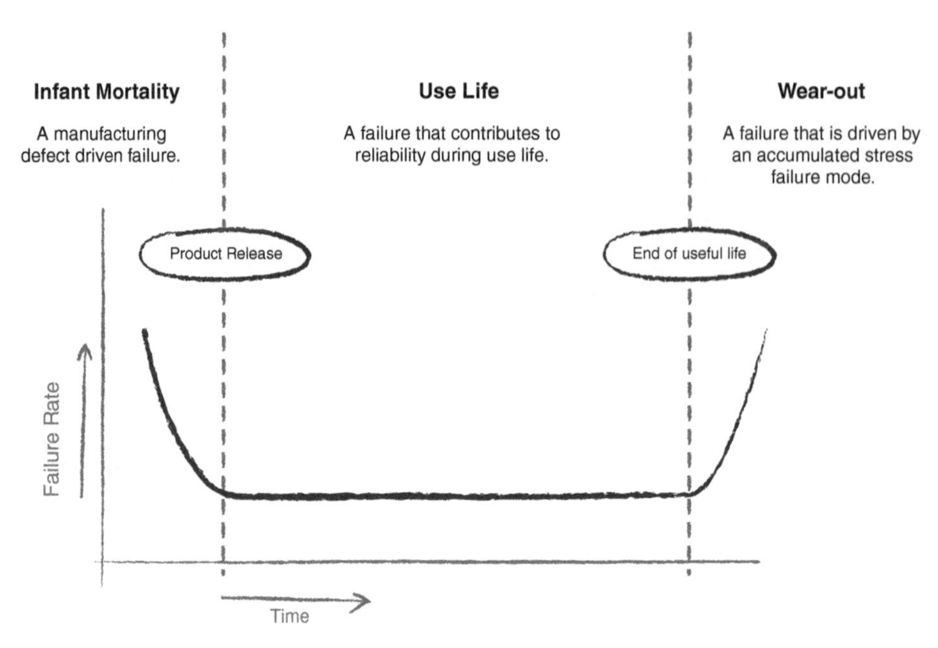

Figure 7.4 Bathtub curve.

Let's take a look at each measure, and why you should pay attention to it. But first let me introduce you to the reliability bathtub curve (Figure 7.4). It is a great visualization for reliability over a product's life.

The bathtub curve demonstrates the failure rate for a population of product through its life. There are three major phases: "infant mortality," "use life," and "wear-out." In the infant mortality phase failures are dominated by quality defect failures. These failures show themselves relatively quickly and those units affected drop out of the population. The use life phase can be most easily characterized by a constant failure rate. This is represented by the reliability we promise the customer during use life. In the wear-out phase the failures are driven by the device wearing out. These are predictable failures that are basically guaranteed to occur and are what happens to a perfect product.

That generic bathtub curve is exaggerated and simplified so it is a good demonstration tool. Figure 7.5 shows a curve that would be more indicative of a well-engineered product with a modern-day quality process.

The two most critical pieces of information on this chart are the failure rate during use life and the time to when wear-out failures become dominant. Failure rate during use life is represented by the height of the bottom of the curve. The time until the wear-out failure modes become dominant is indicated by the vertical divider of the third phase (Figure 7.6).

The Primary Wear-out Failure Mode

The "primary wear-out failure mode" is a doozy. It's potentially tragic. Wear-out happens to just about everything.

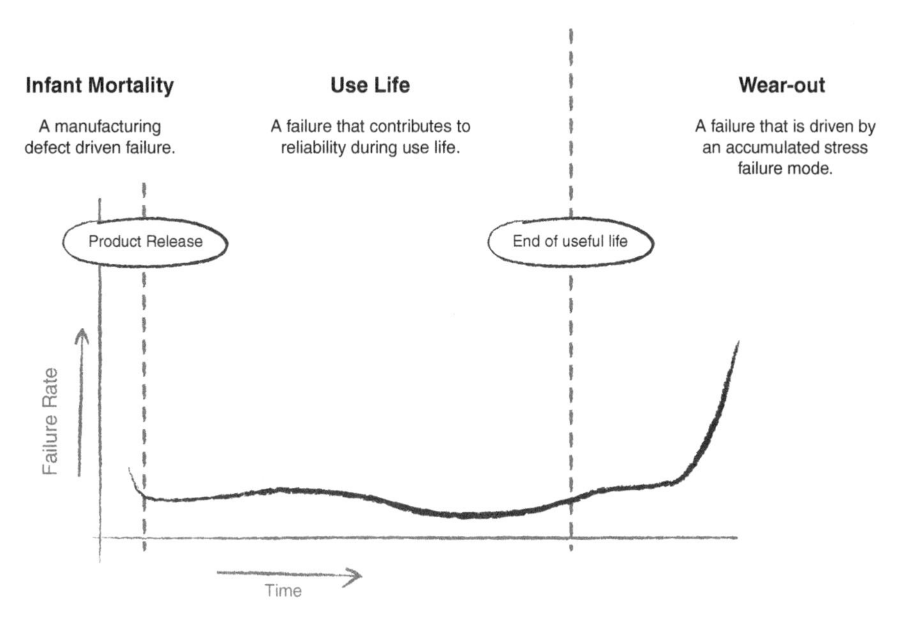

Figure 7.5 Real bathtub curve.

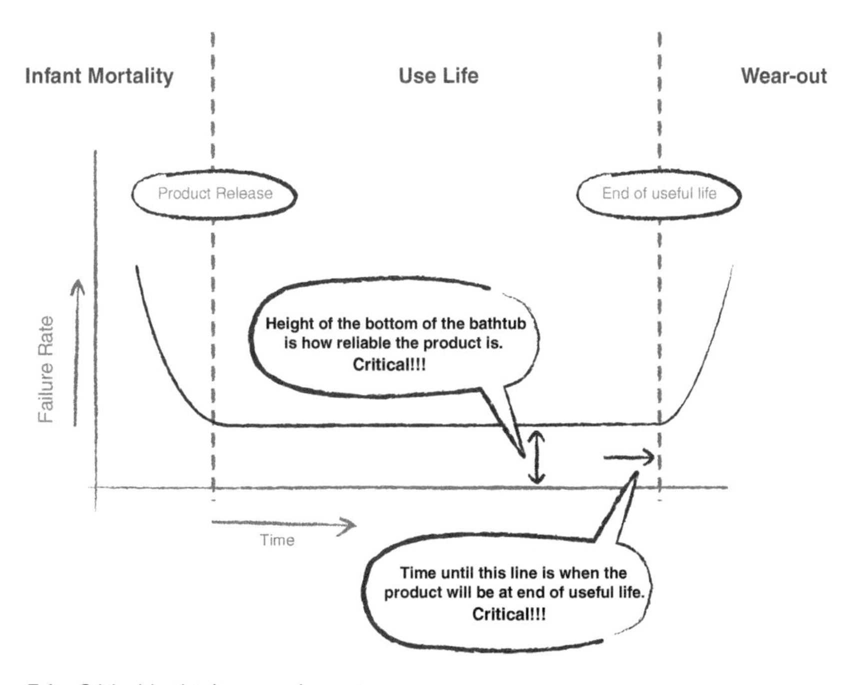

Figure 7.6 Critical bathtub curve elements.

The primary wear-out failure mode tells you when your product will no longer do what it's supposed to do. In other words, if you've manufactured a portable juicing machine, this measure will alert you as to when the motor will stop, or the on-button sticks, or a blade breaks and leaves metal shards in the smoothie.

These things, by the way, aren't failures that are random. They're failures you can expect. They're expected outcomes, based on the materials involved. It's not that you're using substandard parts or manufacturing them in a shoddy way. It's just that everything has a probable lifespan.

So this measure lets you know when these failures will likely happen, so you can change the design in a way that prevents accidents and failures from wear-out.

The Random Fail Rate During Use Life

The "random failure rate" tells you what you can expect in terms of warranty hit. It'll also tell your customers how much they can depend on your product.

The random failure rate is, in fact, what we think of as "reliability" when we think of product reliability. It answers the question, "If I turn this key, will the car start?" or, "If I tell my voice-activated speaker to play a specific tune, will it play that tune or instead give me the weather?"

This is a right or wrong thing. It's a "Is this performing the function I paid for?" thing.

By measuring the random failure rate, you can decide how far you want to take the development. Because development isn't free. The cost is time to market and price point. You get to decide how much reliability to bake in.

Reliability Maturity Assessments

A "reliability maturity assessment" benchmarks an organization's reliability practices against a set of standard definitions. The process includes interviews, review of product performance, and evaluation of best practices. Assessments are carried out on-site, in person, and in depth.

An assessment goes beyond just the engineering aspect of reliability. If we are going to get the whole picture, we have to start at the beginning and continue to find the far-reaching effects of a program. We aren't just covering engineering; we want the whole picture. Individuals from marketing, quality, service, sales, and field service need to be involved. We need to know what they do and what they think about how the reliability process works.

Reliability is a far-reaching discipline. It links into just about every part of the organization. It starts with not just a specific product program plan but also the company's business plan, "What level of reliability does our brand promise?" That's a commitment to your customers, and if broken is not easily mended.

The "why" for any reliability initiative is the same as all other aspects of the product program: money... making it and saving it. The outputs for reliability go beyond the organization. Reliability affects the organization's customers, the customers' customers, and any regulatory organizations they operate under. A complete reliability assessment has to evaluate all of this. This is the complete set of metrics.

How far does the assessment go internally? As far as it needs to. At a minimum it should include engineering, marketing, sales, field service, quality, manufacturing, and the executive staff. I typically set no boundaries and just follow where the clues lead me, like any good detective.

It is not necessary to have a reliability assessment be coordinated with a specific product development program either. Often they get initiated because there is a new product being developed and leadership wants to make sure "They get it right this time." But often that is too late. It can take some time for the recommended changes to take root and produce results. Remember, we are looking to measure and create a path to improvement for the reliability process and everything that interacts with it. It's no small undertaking with a few quick fixes.

Steps for an Assessment

These are the high-level steps for an assessment:

- Select the people to survey.
- Select the survey topics.
- Develop a scoring system.
- Analyze the data.
- Review results with the participants.
- Summarize the results.
- Review recommended actions.
- Assess areas of interest in further detail with related individuals.

Selecting who to survey is important. If you only select people who are available currently or you choose too narrow a cross-section of the team, you're unlikely to gain a true understanding of how reliability functions in the organization.

Selecting who to survey is important. If you only select people who are easily accessible or team leads, you're unlikely to gain a true understanding of how reliability functions in the organization.

Begin the process with a preliminary assessment of the organizational structure. What roles exist and how do they interact with each other and program deliverables? With this information now available, create the list of individuals who should be interviewed first. I always find this is only the beginning. Each interview leads to two other new interviews that get added to the list.

I've conducted surveys where the engineering leaders were certain that the organization's business decisions were based firmly on a product's reliability. Only to discover that the organization's sales and marketing leaders barely paid attention to reliability. Instead, these leaders assumed a high failure rate was par for the course, and that slowing things down by making the product more reliable would be a net loss. A more reliable product that is later to market was only lost sales in their eyes.

In surveys like this, the report often looks like I have assessed two completely different organizations and combined them into one report. One "organization" is completely satisfied with the reliability performance, completely unaware of the trail of disaster in its wake. The other "organization" is exhausted from firefighting and barely keeping regulatory organizations from shutting things down. The look of shock on each of their faces when I share the other's perspective is something to be seen.

This leads us to the other benefit of an assessment, less tangible but equally, or more, of value: connections. Sure, the assessment yielded insights that improved the process. But the benefit in simply creating connections between these groups, which although in the same building might as well have been in different countries, is tremendous.

They not only get the full picture to see who was pushing on the other end in the opposite direction, unknowingly, they also create long-standing communication paths that ensure this distancing does not occur again.

The Team

Some of the roles I try to get on the survey include the following:

- **The design team:** this includes mechanical, electrical, and software engineers, prototyping technicians, chemical engineers, biomedical engineers, and any other technical contributor to the design or base technology.
- **Engineering leadership:** the first- and second-level engineering and science managers are a core part of how reliability affects the day-to-day and long-term strategies.
- **Project managers (PMs):** the PM role varies greatly from organization to organization. I've met PMs who are simply empowered administrators, and I've met others who are more like senior directors and can even call a private meeting with the CEO. Their task is to serve as the program's first-level owner. Their responsibilities include overseeing that the right product is developed, that the program stays on schedule, that the product hits its cost point, and that the program stays in line with the company's objectives and constraints. From this description, you'd think that they'd be the perfect advocate for most of the methodology I'm prescribing. But in actuality they're often under so much pressure from the scheduling goals that all the other objectives fall to the wayside.
- **Manufacturing engineers:** are engaged throughout the full product development process. Their role is to ensure that the design fits today's modern manufacturing processes. They make certain that the ability to easily manufacture and measure is a fundamental part of the design, and not afterthoughts.
- **Quality engineers:** ensure that the product can be manufactured consistently, and that modern quality practices can be implemented while the product is in production, and is being monitored in the field.
- **Executives:** This includes VPs of R&D, quality, service, sales, marketing, as well as the president or CEO. Each of these leaders can bring tremendous resolution into how the reliability of the product and the reliability initiatives affect the business. This is one of the biggest shifts we're trying to make with this new reliability culture. Staying connected to the business. These leaders drive the business. Don't exclude them.

Don't stop there. That is only the starting list. If you believe the person at the receiving dock can offer insight then interview them. You really have to think like a detective.

The doorman may have something that is a key part of the case. To prove my point the receiving dock loader was not a random example.

In one assessment I completed, I found that many people in both quality and engineering didn't know much about the product's journey between the customer complaint and the returned product that arrived in the lab. This product that made it back to the team for root cause investigation was immensely valuable. It held all the clues. I was not satisfied with this gap in knowledge. A lot can happen between "here and there." "Here and there" being any quantifiable distance. It can be one end of a lab to another or a medical tent in Africa to a hospital in Italy. I've seen a thumbprint be a root cause.

So what did receiving know that the rest of us didn't? It turns out that the hub that organized the returns would just throw all of the unpackaged devices into a large cardboard pallet-sized box. The box was then shipped back to the factory. These units were removed from the box and then sorted by serial number.

The quality engineers received a nicely organized set of return items, in serial number order. They had no idea that half the damage "clues" were simply from the units freely bouncing in a large crate as they found their way back to the factory.

The product was basically receiving the worst beating of its life just getting back home. The product was a very delicate electronic device. This is a quote from my summary, "This process is akin to dragging a patient behind the ambulance." The team had been chasing their tail with bad root cause data for years.

The survey topics should be prepared in advance. Yes, I do like to let my findings guide where I go next, but that's no reason to not have a solid plan up front. There should be at least 30 questions prepared. The questions are derived during the pre-evaluation. The pre-evaluation itself doesn't have to be well planned. It's primarily a "survey of the land." By simply talking to one VP for 15 minutes, an engineer for 30 minutes, and a service person in the hallway, much of what needs to be delved into will come into focus.

The Topics

Some categories of questions commonly incorporated are the following:

- "How are reliability goals defined and communicated to the team and leadership?"
- "With a new product development program is risk in the design accessed in advance of distributing the resources?" "How is it assessed?" "Who is included in accessing it?"
- "What are the derivatives of design for "X" (DfX) that you see regularly incorporated into the design process?" A mechanical engineer may list five and a VP of manufacturing may say they don't do DfX. This would be a tremendous red flag as to whether DfM (manufacturing) is really incorporated by the manufacturing engineers or is just something the design engineers think they can do as "experts."

It is evident how some of the greatest reveals come from asking the same question to multiple people. There answers aren't what's most interesting. It's how different their answers are that is very telling about connectivity. It's not that different from asking the kids separately who broke the lamp. The differences in the stories are where the investigation will lead us.

The Scoring

A scoring system is needed to help with translating some of these opinions into quantitative guidance. It can be as simple as:

5 = extremely effective
4 = effective
3 = moderately effective
2 = slightly effective
1 = not effective
0 = not done or discontinued
DK = Don't know.

To me, "Don't know" is the most important value in the scoring scale. Note that "Don't know" and "Not done" are not the same. "Not done" means I can confidently state that we do not do this. "Don't know" means it may or may not be done: the interviewee is unaware.

The scoring system is translating their judgment into a quantitative measure. It's important to emphasize that we are talking about performance, not how often or thoroughly it is applied.

After an assessment the thing that everyone wants to know is, "What should we do first?" After the effectiveness scoring has painted a general landscape of where improvement is, we need to draw a complete landscape using the nonquantitative data to fill in the details.

The first recommendations typically derive from the areas of greatest discrepancies. These discrepancies can be from two far-reaching corners of the organization or between two individuals that sit in the same open concept office. Side note: do "open concept" offices seem to result in everyone pulling a hoodie over their head and wearing headphones? I don't think the "concept" is having the intended effect.

I do notice that the more group communication has become automated and locked to specific avenues, the more information is lost. I have solved more than one mystery by just happening to be at the communal coffee pot at the right time. "I'm sick of receiving all these returned units just thrown in a box. No packaging, nothing! It takes forever for me to sort them out."

Reviewing the results with the participants sometimes yields so much information that I have to go back and rewrite the assessment. The shared "Ah-ha!" moments can yield some amazing information. I guess that's because it's a little similar to all of us being at the communal coffee pot at the same time.

A great deal of new information comes to light as the team realizes how their observations are connected to how other parts of the organization operate.

Analyze: The Reliability Maturity Matrix

A common benchmark used in reliability assessments is the "Reliability Maturity Matrix." It's a five-level guide that characterizes organizational maturity in relation to reliability in seven categories. It's a derivative of the type of analysis created by Philip B. Crosby. In the 1970s, Crosby created the Quality Maturity Matrix as a component of his new Quality Methodology Toolbox [2].

The Reliability Maturity Matrix is divided into five different stages and seven different categories from which to determine the stages of maturity. Read across each row and find the statement that seems most true for your organization. The average of the stages is your organization's overall maturity stage. Quite often, the level of maturity will vary depending on the category, but it is unusual for an organization to vary by more than one or two levels from one measurement category to the next.

It is rare to find companies that are strictly Stage 1. Often when we approach companies with the idea of performing an assessment, their response is, "Why bother? We already know we do absolutely nothing in the way of reliability." Most companies have some form of reliability program, but it usually isn't well documented or isn't well practiced. Companies usually have pockets of excellence within their organization that we can extract and highlight.

It is also rare to find companies that are strictly Stage 5. Stage 5 is reserved for "best in class" companies, and there aren't too many of these (and for the ones that do exist, they probably won't share their techniques with the rest of the world, because this is a competitive advantage). Stage 5 companies have their processes down so well that they don't need assessments to tell them where they are and typically don't call in outside consulting teams to assess their processes – they already know where they are.

Figures 7.7 and 7.8 show the five-stage table and the definitions associated with the seven categories.

Review with the Team and Summarize

Have the team review the content and aid in the process of developing recommended actions. There is a good deal of additional knowledge and observations that can be extracted in this process. But that will only occur if it is a real part of the process. What I mean by "real" is that it is not just a check in the box, similar to reviewing a clause in a new phone contract or software agreement.

If the review is a mass email inviting feedback then it is nothing more than that check in the box. That email does two things:

- It gives the impression that included information is mostly finalized, not inclusive.
- It also implies that the topic is low priority and only deserves a passing glance.

The review should be a set of meetings, even if they are short. By holding at least two short (<1 hour) meetings the team is again forced into discussion about the analysis. What will occur is a summary that drives specific actions. These aren't just prescribed actions; these are actions the team has agreed to as helpful.

It didn't exist if it wasn't documented. I also say it didn't exist if no one wants to read it. There should be not only a clear record of all the work but also a concise summary. For the analysis to be effective it must be communicated and understood outside of the group that participated in it. For this to happen we need a clear summary that can be easily shared.

Select a summary format that fits the organizational culture. If this is an organization that transfers significant results, and initiatives in team meetings, then make it a presentation. If this organization uses an online virtual team software package to communicate, make sure it gets uploaded to that platform.

Measurement Category	Stage I: Uncertainty	Stage II: Awakening	Stage III: Enlightenment	Stage IV: Wisdom	Stage V: Certainty
Management Understanding and Attitude	Management has no comprehension of reliability as a management technique. Management blames reliability engineering for "reliability problems".	Management recognizes that reliability may be valuable, but they aren't willing to provide much money or time to make it happen.	Management is still learning more about reliability. They are becoming more supportive and helpful.	Management participates and understands the absolutes of reliability. They recognize their personal role in continuing emphasis.	Management considers reliability an essential part of company system.
Reliability within the Organizational Chart	Reliability consists of a single engineer who may be doubling as a quality engineer or another function. Reliability is hidden in manufacturing or engineering departments.	A stronger reliability leader is appointed, yet the reliability function is still buried within manufacturing or engineering departments.	Reliability manager reports to top management with a role in management of division.	Reliability manager is an officer of the company and is involved with consumer affairs.	Reliability manager is on the board of directors. Reliability is a thought leader.
Problem handling	Organization is in firefighting mode; no root cause analysis or resolution takes place.	Teams are set up to solve major problems. Long-range solutions aren't identified or implemented.	Corrective action process is in place. Problems are recognized and solved in an orderly way.	Problems are identified early in their development. All functions are open to suggestions and	Except in the most unusual cases, problems are prevented.

Figure 7.7 Maturity matrix (part 1 of 2).

Measurement Category	Stage I: Uncertainty	Stage II: Awakening	Stage III: Enlightenment	Stage IV: Wisdom	Stage V: Certainty
Cost of Reliability as % of net revenue	Warranty: unknown Reported: unknown Actual: about 20%	Warranty: 4% Reported: unknown Actual: about 18%	Warranty: 3% Reported: 8% Actual: about 12%	Warranty: 2% Reported: 6.5% Actual: about 8%	Warranty: 1.5% Reported: 3% Actual: about 3%
Feedback process	No reliability testing is performed. No field failure reporting other than customer complaints and returns.	Some understanding of field failures and complaints. Designer engineers and manufacturing don't get meaningful information.	Reliability manager reports to top management with a role in management of division.	Refinement of testing systems -- only testing critical or uncertain areas. Increased understanding of causes of failure allows deterministic failure rate prediction models.	The few field failures are fully analyzed, and product designs or procurement specifications are altered. Reliability testing is performed to augment reliability models.
DFR program status	No organized activities occur. Organization has no understanding of such activities.	Organization is told reliability is important. DFR techniques and processes are inconsistently applied and only occur when time permits.	Implementation of DFR program with thorough understanding and establishment of each technique.	DFR program is active in all areas of company -- not just design & manufacturing. DFR is a normal part of research and development and manufacturing.	Reliability improvement is a normal and continued activity.
Summation of reliability posture (actual quotes from companies)	"We don't know why we have problems with reliability."	"It is absolutely necessary to always have problems with reliability?"	"Through commitment and reliability improvement, we are identifying and resolving our problems."	"Failure prevention is a routine part of our operation."	"We know why we don't have problems with reliability."

Figure 7.8 Maturity matrix (part 2 of 2).

The trick is to make sure there is follow-up discussion, no matter how casual, with the leaders we need to be aware of this analysis. Casual follow-ups with a lunch meeting or a formal discussion with them and their team are both sufficient. The important factor is that there was a point in time when the leader of the analysis and those critical leaders directly connected.

Recommend Actions

After you get the feedback and feel confident that you have captured the essence of an organization, it is time to come up with recommendations. This is perhaps the most difficult portion of the assessment. This is where experience really counts. You should have a good understanding of "best practices" in the industry to be able to discover a pattern and draw a conclusion from the pattern. If you have never performed an assessment before, I recommend that you call in an expert to help you with your first assessment to make sure you perform it thoroughly and to make sure you draw the proper conclusions. From the assessment, you are looking for trends, gaps in processes, skill mismatches, over-analysis, and under-analysis. Look for differences across the organization, pockets of excellence, areas with good results, and areas that need work.

No one technique or set of techniques makes an entire reliability program. The techniques must match the needs of the products and the culture.

Many companies score a Stage 2 or Stage 3 and then ask what they need to do to reach Stage 5. First of all, achieving Stage 5 is quite rare. Secondly, moving more than one stage within one product release is also rare. You should set your expectations appropriately and be patient while you change your systems to achieve better reliability. If you try to make changes too quickly, it is likely that the changes will be rejected by your team or your reliability program will start breaking down.

Assess Particular Areas in More Detail

During the assessment, there are bound to be areas that require further investigation. A list of survey topics may not be enough to assess the situation and provide recommended feedback. For these areas, we recommend a more detailed assessment. Some of these areas may be:

- Review of field results.
- Review of manufacturing yields.
- Assessment of key supplier(s).
- Review of design control process.

You can then add the results from these detailed assessments to your overall recommended actions.

Golden Nuggets

"Golden Nuggets" refers to those few techniques that your organization does well, so well in fact that these techniques become engrained into your culture. The Golden Nuggets become part of your "secret sauce" that give your product and company a competitive advantage in specific areas. Sometimes your organization doesn't know that they do them well or even know that they are doing them at all. It is your responsibility to point out these Golden Nuggets to the organization because you should always reinforce good behavior.

When assessing your organization, look for these Golden Nuggets. Perhaps your organization has a keen ability to use mechanical simulation techniques such as Finite Element Analysis (FEA) or statistical data analysis techniques such as Design of Experiments (DOE). If you discover this, make sure to point it out to the organization and help them use these Golden Nuggets to their advantage.

Other times, you may discover that your organization has an excellent ability to work as a team and to reach a consensus. Perhaps they don't even recognize that they possess this skill. Again, you should encourage this behavior when you recognize it.

Summary

It really is amazing how simply measuring something sparks improvement. It's human nature. What isn't being measured in your organization that is, in fact, an important factor in your operation? Just list five right now. What would it take to measure them? Could it be as simple as asking an individual with a role that has access to the pertinent data to just. . . measure it?

Having a well-defined (and well-followed) product development process could be another example of a Golden Nugget. Many times we have seen companies with disjointed product development processes, and when we followed failures back to a root cause, we often found that they bypassed a process or process step. Consistently following a product development process is a key part to reducing failures, and the companies we have worked with that already had this in place were typically the companies with more reliable products.

References

1. Drucker, P.F. (1955). *The Practice of Management*. Oxford: Butterworth-Heinemann.
2. Crosby, P. (1979). *Quality is Free: The Art of Making Quality Certain*. New York: McGraw-Hill.

8

Reliability Culture Tools

Advancing Culture

This section of the book is for the engineers, managers, and leaders alike in product design who want to learn the tools that make reliability culture happen.

In this particular chapter, each section presents a fundamental technique for shifting your culture more strongly toward reliability. If you apply these techniques sincerely, the cost and time investments required to make great gains will be minimal.

Notice I used the word "sincerely." The reason: we often see methodologies applied in a manner I can only describe as "insincere." They're learned, box-checked in some development plan, and applied . . . once. Shortly thereafter they fade as we revert to the familiar.

If you see this kind of fade happen, call it out. After all, most people don't revert to previous behaviors for evil reasons. It's just something that happens when well-intended people who are under pressure try to do too much in too little time. Keeping up the attention required to make a change stick at times is too much to ask.

What happened was that the organization wasn't genuine in their desire to make the change. They didn't allow their leaders the bandwidth to keep the new process on track.

The methodology I'm proposing is called "Reliability Bounding." It keeps your most cherished reliability goals front and center at all times, and requires minimal management to make them stick. It also keeps those goals in the proportions that they were originally set, so they guide your daily program actions accurately.

Where did these goals and proportions come from? Your business and marketing plan, which has one objective: to gain and hold market share.

The Bounding tools ensure that your program, reliability plan, and business objectives stay aligned. Alignment doesn't just increase efficiency. It reduces human conflict and the need for manipulative managing.

Manipulative Managing

What do I mean by manipulative managing?

In simple terms, it's when a leader guides their team through indirect means. In other words, instead of coming out with what they want directly, they only allude to it.

Reliability Culture: How Leaders Build Organizations that Create Reliable Products, First Edition.
Adam P. Bahret
© 2021 John Wiley & Sons Ltd. Published 2021 by John Wiley & Sons Ltd.

If leadership uses a tactic like that, it's usually based on two factors. The first is that the staff's personal goals and the company's goals aren't completely inline. For instance, a salesperson receives a large quarterly commission based on sales alone. The problem: that commission doesn't consider if their customers are satisfied. So, the salesperson gets their bonus and is satisfied, while the company may be stuck with angry customers and is dissatisfied.

The second factor is that the leader feels their directives are being met with strong resistance.

Manipulative Management in Action

Here it is manipulative managing in action.

I worked with a CEO who wanted counsel on how to improve his program reliability. Along with my organizational evaluation, he was open to my evaluation of his leadership strategies. That he was open to change was fortunate, because his take on leadership left a lot to be desired.

My first observation was that he led his team by creating fabricated boundaries. The team often had obstacles that looked legitimate on the surface, but upon closer examination weren't. I saw the CEO create more than one conflict intentionally. At first I couldn't understand why.

There were two initiatives to complete a design. That's pretty common when there are multiple paths being explored in a new technology program. The CEO cut the budget on both initiatives, and the reasons he gave were vague. Something about overspending elsewhere and needing to balance the budget.

The result? The team leaders began poaching team members from each other, and began stealing resources for their own project. Why would the CEO give an edict that created such aggression and chaos?

He had decided that "Initiative B" should stop. It simply wasn't going to serve the program well. He wanted to put all the resources behind "Initiative A."

B, though, had supporters. One was a senior person on the CEO's executive staff. He feared that if he used his authority to cut B the team disappointment would cause program harm.

So what happened? There was so much conflict between the teams from the budget cuts that the CEO had reason to step in and break it up. He immediately cancelled B and put the blame on both teams for not being more collaborative. He did this when team B had thrown the most recent punch. This way without even having to say it the implication was that B was the one to have broken the process.

The CEO got what he wanted and he wasn't the bad guy. He actually had the leaders from team B feeling remorseful. The result was not only being able to shut down B without blow back but also having the team that was shutdown supporting A as a means of mitigating reprimands. A tactic right out of Machiavelli's *The Prince*, if I ever saw one.

An Alternative to Manipulation

In the end, great results. But give me a break! What a rigmarole, and a waste of time and energy. Why wasn't there a way for the CEO to have his entire staff see the benefits of stopping "B" and fully funding "A"?

He never tried to get the team to understand the "why." Why couldn't both initiatives continue in parallel? Why was it best to choose A?

If the teams were aligned with the big picture and the "why" was understood, the CEO could have simply let the obvious unfold. Team B would have agreed that their initiative was not going to be successful. But they had too much personally wrapped up in its success to let it go.

What makes a direct method work is that the receiver understands and agrees with the "why." It was the "why" he believed he had to fabricate. The Bounding method was what I advised to the CEO so going forward he could lead in a straightforward manner.

Transfer Why

The Bounding method aims to transfer the company's objectives to the team.

Now, of course, the team already knew the company's objectives, but did they own them?

The team had personal objectives associated with their roles. This is what guided them day to day. "I want that raise." "I want that promotion and new office." This was solely what guided their actions. It wasn't that these people were narcissists; it was how the system was structured. This is why the CEO had to manipulate them. He was correcting for a system that was ineffective.

So there is a bit of irony if we consider that he was one of the architects of the system they operated in. He created the role structure and incentives. CEOs measure project managers on time to market. CEOs measure R&D teams on the success of new technology.

Reliability Bounding

Reliability Bounding is based on a single principle: there are multiple objectives in any program and they should all drive the day to day.

When the highest-level goals are not providing these day-to-day inputs, something else is. That "something" is usually everyone's personal goals, actual or perceived, and unforeseen outside forces. I call this type of plan execution "fire and forget."

Fire and Forget

"Fire and forget" is how an old-style rocket works. With fire and forget, we point something in the desired direction, pull the trigger, and hope for two things: that the trajectory was calculated correctly, and that nothing unexpected happens. If the target elevation was wrong, well, nothing you can do about it. If the wind kicks up, just cross your fingers and hope for the best.

The rocket is your team barreling along out there with no further guidance. Bounding ensures that the target is driving corrections during the entire journey.

Reliability goals are one part of the target we want to hit. We have discussed the others (time to market, cost point, and new features). There are many ways to get that closed loop control with those factors. For right now I will focus on reliability. You'll see how the method can be applied to other goals as we review it.

Reliability Feedback

For reliability engineering in product development, what feedback are we looking for? How do we know we are on track? Reliability engineering has to generate this guiding information. Rarely can it be found elsewhere. There are no readymade feedback loops for reliability. Time to market has an automatic feedback loop and rate of progress, and a desk calendar covers it.

A good reliability program immediately puts into motion the test methods that aim to "measure" percentage in availability – measure confidence in the reliability goal, measure how robust the product is, and measure when the full population is going to wear out.

So why do the people who pay for these tests not read the report? Or even care that the report is going to be arriving a year after the product left the factory? At that point, skip the testing and just ask the customer how the product is doing.

The Reliability Bounding process is in two distinct phases. The first is "Strategy Bounding," which lays the foundation, guiding factors. The second is "Guidance Bounding," which creates the closed loop control when the program is in motion.

Strategy Bounding

Strategy Bounding is a part of program planning. The program will hit its reliability targets through architecture. The process starts with evaluating candidate reliability tools for best return on investment (ROI).

ROI is difficult to quantify for reliability tools. A good deal of their value is through mitigation of future issues. These issues might even occur when the current team is disbanded.

Other returns, like "customer satisfaction" and "contribution to future sales," are even more difficult. These can become completely detached if we don't find ways to tie them in.

By using ROI estimations for tools we can select the ones that deserve the most resources. This evaluation done early will do something else very important: it will be the lens we use to evaluate new information throughout the program.

Strategy Bounding Toolkit

In Strategy Bounding we will use the following tools:

- Bounding ROI
- Anchoring
- Focus Rotation.

The targets for product factors that are established to guide the product development program will guide the Reliability Bounding process as well. Many product factors influence a program. It is important that they relate to the program's business objectives as well as be firmly established at the program start. They will likely be renegotiated and adjusted as the program unfolds. As I outlined in earlier chapters, the four most common product factors are:

1) Time to market
2) Product cost point
3) Target reliability
4) Features.

These four factors, with the addition of the program cost, are the guiding forces. They affect major decisions for the duration of the process. The big questions that are often so difficult to answer are "How do you measure the effect of a decision on each of those factors?" and "How do you negotiate the value and effect on each factor relative to one another?"

Midprogram Feedback

Come midprogram, you may get questions that you couldn't have planned for at the start: "Do the three additional months requested for the Accelerated Life Test (ALT) completion justify the later time to market?" "What do we do when we don't have a high confidence the product is going to last as long as promised?" "Should we release it late which may cost market share just to do more testing for increased confidence in long life?"

Historically, these decisions are made ambiguously. Mostly based on who yelled the loudest at the Friday steering meeting. Or, more importantly, who can make the biggest case that their scenario leads to better market share, whether it's quantifiable or not.

If these discussions are occurring without all the facts, which they often do, we're basically just steered by emotion, debate, and power dynamics. None of these leads to actions that best serve the program, product goals, or the company.

Our first task in applying the Bounding method is to find a way to quantitatively relate the impact and benefit of each of the factors to the program. The debate as to which issue, or factor, is more important has to be concluded in the program early. Changing these factors midprogram can be very harmful to the process in terms of delays and expense. The scoring and relationships we will define are relative and not absolute. A change in any values will cascade to all previous and future decisions if the relative positions get out of sequence.

The Bounding Number

So what's this correlation factor called, which is the foundation of the Bounding tool? Ingeniously, I call it "the Bounding number." (I never said I was creative at naming. My first choice for naming our kids was "Thing 1" and "Thing 2," practical and easy to remember, and courtesy of Dr Seuss!)

I'll explain what a Bounding number is, and how you'd use it in your reliability program, by first talking about planning a vacation.

Suppose you wanted to go on a vacation. How would you know if your vacation could last one week or two? The answer (or at least the simplified answer in this case) would be money. Money is the unit of measure that allows you to think about a lot of disparate things strategically. It's a unifying factor that helps you rearrange aspects of your life in a way that helps you.

To go on vacation, you need money. What's the right amount? You have to consider your expenses. Each month you have a rent or mortgage budget, a food budget, a clothing budget, an entertainment budget, a car maintenance budget, and so forth.

Even though these expenses affect very different parts of your life – I mean, your rent payment and your entertainment payment are two very different things; one's a necessity while the other is a nicety – you can still bring all these different parts of your life together by correlating them with money.

So, should your vacation last one week or two? Well, you have to look at your various budgets and see what you can swing. Maybe you'll cut out buying new clothes for a few months. Maybe you'll curtail eating in restaurants. Maybe you'll see if you can combine your Internet, TV, and phone payments with one provider so you can save costs.

Suddenly, things are brighter. You're looking at parts of your life that have very little to do with your vacation, like what you eat on a given day, and you're using them to help finance a vacation of two weeks.

That's a major benefit of money. It helps you reconfigure your resources, so you can think about the different parts of your life clearly and budget for your needs as they change.

The Bounding number, then, acts as money for your product development program.

A leader wants the organization's new product to feature the latest technology, sport a competitive price, be highly reliable, and get to market faster than anyone else's.

But each one of those things comes at a price. You can't do them all to the nth degree simultaneously. There'll be tradeoffs. There must be tradeoffs. For instance, you can't release your product early and have enough time to do a full, comprehensive test program. Those two things are at odds with each other.

How do you decide which tradeoffs you'll accept and which you won't? Your Bounding number will help you decide. It'll allow you to weigh your options, one against the other. That way, you won't be making important product decisions based on your gut alone. You'll have a kind of data measure that will help you make your decision more objectively.

Bounding ROI

OK, onto the Bounding methodology.

To begin, you'll need to create a numbering system that helps you keep track of your product program investments.

By the way, when I use the term "investments," I'm not talking about only money. I'm also talking about the program's calendar time, machinery wear-and-tear, work hours, time to market, and so forth.

The numbering system takes everything that could help or hurt the product release and your company's reputation, and puts it on a scale. The scale comes from your knowledge and judgment.

It's like when Amazon asks you to rate one of its products on a scale of one to five stars. You'd inherently know the difference between a one-star rating (bad) and a five-star rating (great). To arrive at the proper star rating, you don't need to do much of a calculation. You don't have to watch any explanatory videos on how to rate a product. You just know.

And a related point about Amazon's rating system: while they chose a one-to-five star system, they could have just as easily have created a system of one-to-a-hundred stars. Why one system over the other?

The one-to-a-hundred system would have been too much. It would have made you do too much thinking. You'd likely not leave a rating at all, because it would have been too much work. It would have hurt your engagement with the site. Instead, they decided to keep things as simple as they could, while still providing a meaningful delineation between items being rated.

Now keep those ideas in mind as you select your program's Bounding system. Selecting your numerical range is simply a matter of how much resolution you believe you need.

For most projects, I find a scale of 0–25 does the job.

You'll see in the following tables what that value is for the different parameters.

To put the Bounding method in motion we need established Bounding values. These are the numbers we use to relate the value of investments and returns.

Remember, the Bounding value is like a currency. The same way money links your hard work (investment) to getting things you desire (returns).

What can you buy with 20 hours of your time? Well, because of currency it is simple to figure that out. We convert both your work and the item you desire into dollars. We have to set conversion rates for each.

We request $35 for each hour we are fixing cars. We are working extra hours because we want a huge new TV. We have big plans for Saturday nights with friends and family.

We look online and see that the price on that big beautiful TV is $980. Looks like we have 28 hours of overtime ahead of us. So there you go, we know exactly what amount of investment will get us what we want in return.

The range of the Bounding scale is up to you. The only requirement is that it is consistent across all Bounding tables. You can't have 1–10 in one table and 1–25 in another.

If we've selected 0–25, a value of 0 as an investment means, you guessed it, nothing is invested in the program. A zero in return means nothing was gained by the investment. In our work analogy, zero work represents zero dollars, zero dollars means you don't get stuff. Twenty five would represent you spending all of your money. How much you get is dependent on what the 25 equates to. You will see in the following tables the selected values for investment and returns in reliability.

Invest and Return Tables

Let's make a scoring table for reliability investments and returns. The factors of investment in reliability tools is either financial or in added time to the schedule, or both. The first investment we will score is "time." We will say that investing three months in a testing program is equivalent to a Bounding number of 5. The most we would expect to invest in a reliability tool is 15 months. We'll put that at the top of our Bounding scale at 25. We now have a transferable proverbial currency for time invested in reliability. Three months is 5 Bounding units and 15 months is 25 Bounding units.

How about for financial investment? We will set a $100k investment equivalent to a Bounding number of 5 and a $500k investment equivalent to a Bounding number of 25.

If an ALT test will add three months to the program schedule and cost $100k, that is a total investment of 10 Bounding units ($5+5=10$).

How about Bounding numbers on "returns"? We need a way to measure what we get back, "What do I get for my three months of delayed product release and extra $100k added to the budget?

The four things I often identify as returns for reliability investment are:

- Mitigated warranty cost and field fix expense.
- Mitigated loss of future sales.
- Mitigated brand damage.
- Mitigated upcoming project delay due to resource diversion.

These are the typical things we want in return for our time and financial investment in reliability. This is the TV, car, clothing, and stereo we are thinking about while we are sweating out there with cut knuckles and grease on our face while working on a customer's car.

We use the same 0–25 scale. (let's abbreviate Bounding number to "B#"):

- Mitigates warranty range is (\$30k= $5_{B\#}$: \$1mn = $25_{B\#}$).
- Mitigate sales loss range is (\$40k = $5_{B\#}$: \$400k = $25_{B\#}$).
- Mitigate brand damage range is (5% market share loss = $5_{B\#}$: 35% future market share loss = $25_{B\#}$).
- Mitigate upcoming project delay range is (Short delay = $5_{B\#}$: Cancelled = $25_{B\#}$).

Figure 8.1 shows a full set of Bounding conversion tables.

With these tables completed we are ready to use the Bounding method. The tables will help us select reliability tests and analysis tools for the program. We no longer have to do that task by "gut feel." These tables will serve an important additional purpose midprogram. They will assist with evaluating if initiatives should be continued. This can be a big deal when program phase gate freezes are approaching. We'll cover how to do that in a later section.

Let's set the stage. We are starting our program Do we want to do a reliability growth (RG) test program? They are very expensive in both dollars and added program time. But they can save you from a product release disaster. Let's evaluate what investments in time and money will be needed first.

Adding an RG test program will add 12 months to the program schedule. It looks like we would spend about \$200k on RG when all is said and done. So, as per our Bounding table, the conversion would be:

- A 12-month program extension for RG is $20_{B\#i}$.
- An RG program will be \$200k, so $10_{B\#i}$.

You might have noticed something there. I added an "i" to the Bounding symbol. This designates it is an "invest" Bounding number. An "r" will be added for return Bounding numbers.

How about the return values for RG?

- Mitigated warranty cost and field fix is \$110k, $1_{B\#r.}$
- Mitigated sales loss is \$80k, $10_{B\#r.}$
- Protection of damage to brand is 5% market share, $5_{B\#r.}$
- Mitigation of resource loss on upcoming products is short delay, $5_{B\#r.}$

Figure 8.2 shows this in summary.

Bounding Conversion Tables

Invest: **Program Schedule Add**	
Months	Bound #
3	5
6	10
9	15
12	20
15	25

Return: **Mitigated Warranty, Investigation, Field Fix**	
Months	Bound #
3	5
6	10
9	15
12	20
15	25

Return: **Mitigated Sales Loss**		
% of projected max year sales	$	Bound #
5%	$40K	5
10%	$80K	10
30%	$240K	15
40%	$320K	20
50%	$400K	25

Invest: **Program Schedule Add**		
% of program budget	$	Bound #
5%	$100K	5
10%	$200K	10
15%	$300K	15
20%	$400K	20
25%	$500K	25

Return: **Mitigated Brand Damage**	
% Future Market Share	Bound #
5%	5
8%	10
12%	15
20%	20
35%	25

Return: **Mitigated Future Project Damage**	
Level	Bound #
Short delay	5
Long delay	10
Objectives include legacy reliability improvement	15
Product focus is legacy reliability improvement	20
Cancelled	25

Figure 8.1 Bounding return and investment tables.

Assembly: Robot Z **Tool: Reliability Growth**	Value	Bound #
Invest Schedule Add	12 months	Invest 20
Invest: Cost Add:	$200K	Invest 10
Mitigated Warranty & Field Fix	$110K	Return 12
Mitigated Sales loss	$80K	Return 10
Mitigated Brand Damage	5%	Return 5
Mitigated Future Project Damage	Short	Return 5

Figure 8.2 RG Bounding table.

Let's do this for a Design Failure Mode Effects Analysis (DFMEA), an ALT test, and a Highly Accelerated Life Testing (HALT) test as well. Figure 8.3 shows the tables for all four of those proposed program reliability activities.

Deciding by Bounding

Bounding can now help us decide which reliability activities we should include in our program. For the first time instead of going with our "gut feel," we can calculate which tools are the best investment.

We need to do that calculation then. Don't worry: it's so simple you could do it on a napkin, at a bar, after a few shots (not recommended, though). We simply sum the investment Bounding numbers and compare that to the sum of the return Bounding numbers. Whichever is greater directs our decision. If we get more than we invested, it's a good move. If we get less than we invested then we may want to skip it.

In the case of RG our investment total is $20_{B\#i} + 10_{B\#i} = 30_{B\#i}$ Return is $12_{B\#r} + 10_{B\#r} + 5_{B\#r} + 5_{B\#r} = 32_{B\#r}$. The return is greater $32 > 30$. RG is a smart move for this program. Does the gap between the numbers matter? Definitely. In the case of this RG program, the return was only two points higher than the investment. If things get tight midprogram, there is a good chance some concessions may be made. But right now it's clear we can benefit by including it.

In an upcoming section I will discuss using the Bounding number to resolve resource issues midprogram. But for now, let's discuss "Anchoring," a way to keep things connected.

Anchoring

Reliability test and analysis tools are used to either measure or improve reliability. We include them because we want some kind of results. When the program is complete, we look back to see how well they delivered. I think we can agree that this is often disappointing. Why do we so often not get what we expected? There are a few reasons but one of the most significant is "loss of synchronicity." Maybe that seems like a strange term, but it means exactly what it says. We were coordinated once, but now we are not. A day late and a dollar short, so to speak. Actually it is literal, we do end up late and underfunded.

Assembly: Robot Z **Tool: Reliability Growth**	Value	Bound #
Invest Schedule Add	12 months	Invest 20
Invest: Cost Add:	$200K	Invest 10
Mitigated Warranty & Field Fix	$110K	Return 12
Mitigated Sales loss	$80K	Return 10
Mitigated Brand Damage	5%	Return 5
Mitigated Future Project Damage	Short Delay	Return 5

Assembly: Robot Z **Tool: DFMEA**	Value	Bound #
Invest Schedule Add	3 months	Invest 5
Invest: Cost Add:	$50K	Invest 3
Mitigated Warranty & Field Fix	$400K	Return 17
Mitigated Sales loss	10%	Return 10
Mitigated Brand Damage	8%	Return 10
Mitigated Future Project Damage	Short Delay	Return 5

Assembly: Robot Z **Tool: ALT**	Value	Bound #
Invest Schedule Add	3 months	Invest 5
Invest: Cost Add:	$40K	Invest 2
Mitigated Warranty & Field Fix	$5K	Return 1
Mitigated Sales loss	10%	Return 10
Mitigated Brand Damage	3%	Return 3
Mitigated Future Project Damage	No Delay	Return 0

Assembly: Robot Z **Tool: HALT**	Value	Bound #
Invest Schedule Add	1 month	Invest 1
Invest: Cost Add:	$20K	Invest 1
Mitigated Warranty & Field Fix	$200K	Return 15
Mitigated Sales loss	10%	Return 10
Mitigated Brand Damage	12 Months	Return 15
Mitigated Future Project Damage	No Delay	Return 0

Figure 8.3 Tools Bounding tables.

Completing a HALT test at the same time the manufacturing prototype is ready does not serve the program well. We should have had that information at first prototype if it was to be useful. We can't do anything with test results at manufacturing prototype. A manufacturing prototype exists because the team has agreed the design is done and we are looking for ways to tweak the manufacturing process.

The primary value of HALT testing is design feedback to improve robustness. The project manager is going to laugh at you suggesting fundamental design changes so close to field release. So what happened? Why are great design ideas showing up now? We planned HALT testing well. We should have had design input much earlier.

This situation isn't just messy and a waste of resources: it destroys credibility. What's going to happen when you ask for a HALT test budget in the next program?

What would have been good for leadership to know throughout the program is that the value of HALT testing was progressively becoming less valuable. With that information a leader can decide much earlier if they should just cut it loose or maybe increase its priority so it can still add value early on. Why end up with the worst-case scenario: full investment and no return.

The technique that can ensure that this does not happen is "Anchoring". I created Anchoring because I couldn't stand to see this same situation play out over and over again, program after program. It is totally avoidable. As an engineer what we needed was crystal clear: closed loop control.

Closed Loop Control

What's that? It may sound super complicated, but it is not. It's occurring around you every day. In fact, you are the one implementing it. Closed loop control is simply this: you look through the windshield when you drive and use that information to make decisions on what to do next. You observe the road, decide you should go more left or right, adjusting the steering wheel, the angle of the tires change, the car changes direction, you observe where the car is now headed, adjust the wheel again, the tire angle changes again, and on and on until you get to your destination.

Open Loop Control

Try steering a vehicle solely based on a memorized sequence of left and right turns from the map directions, never looking out the windshield. That's called "open loop control." Driving is definitely a bad application for it. Open loop control is more suited for programming a TV remote control. But even that stinks and could benefit from a feedback loop.

So why are we implementing reliability tools in an open loop control style. Programs change. Actually, that's about the only thing you can guarantee about a product program, "It's going to change." Everything else is up for grabs.

Closed loop control systems are complicated to create. A lot goes into creating automated driving systems for vehicles. A lot went into even the simpler version with a human in the loop, behind the wheel. But once closed loop control systems are created they are easy to operate. Actually, that is the entire point of them.

That is what the Anchoring methodology is: a pre-made program closed loop control tool. Just hop in the driver's seat. There are two separate anchors in the method, the Intent Anchor and the Delivery Anchor. We'll discuss the Intent Anchor first.

Intent Anchor

The Intent Anchor is the "why." Why are we using this tool? Why are we spending this money? Why are we extending the program? Without constantly checking in on the "why," we can easily just keep going past our goal or stopping short of it.

How do we identify intent? Why are we doing this test? Why are we investing in this tool? Risk analysis is a great identifier for intent. My favorite risk analysis tool is Failure Mode Effects Analysis (FMEA). The output of an FMEA is a list of risks to the product's success. But it's not just the risks. They are listed in order of greatest to least. That's a big deal! We have limited program resources and we need to know where we should apply them first.

FMEAs don't just output the risks; they include mitigative actions. Many of these risk-driven actions are reliability tools. So the reliability tools on our program list have a clearly identified intent. The problem was that it was last seen in the FMEA notes in the "Next Actions" column eight months ago. Those actions would have remained visible longer if they were written in powdered donut dust on the conference room table.

We need to keep the "why" connected to the "what."

Intent Anchor Statement

The Intent Anchor should always be fully documented in the reliability program plan. Now it's locked in and relevant. As I have said before, the reason we do reliability testing is to either measure or improve reliability. That is why a specific test was selected to address that specific risk. This is an "Intent Anchor statement."

Want an example of an Intent Anchor statement? Here's a good one, followed by a bad one:

Good Intent Anchor statement:

> DFMEA line item #245 dated 4/6/19 identified that wear-out of the new plastic bushing was a high risk for premature end of life of the product. An ALT test will be executed so a statistical statement on the bushing wear-out distribution can be made. If the selected bushing material is found to have insufficient life, alternate designs will be selected and tested. Suggested changes are to be submitted by phase gate #3.

Bad Intent Anchor statement:

> We will generate an MTBF (mean time between failure) number for the full product before release with Life Cycle Testing.

In the good statement the "why" is clear. It answers the question, "Why are we spending money on this?" There is a second critical piece of information there as well: when. We see that the full value of the test is realized if it is complete before phase gate #3.

The bad statement is going to fall apart when challenged in budget or schedule negotiations. There is not a clear "why." It's just a request from an unknown source for an unknown reason.

Delivery Anchor

The output of a tool must be connected as well. We can't just start work and then hope we happen to complete when it is most beneficial. This planned connection between the tool output and the program is the Delivery Anchor. This Anchor is the reference for assessing the tool's value decay, like an expiration date stamped on top of a yogurt. The closer we get to, or farther we get past, that date the less likely that yogurt is going to be satisfying. It anchors our reference for freshness (value).

The most important word when discussing the Delivery Anchor is "decay." Traditionally we treat the delivery of test and analysis tools as helpful or not, nothing in between, totally binary. We aren't catching a train here, when getting to the platform before 2:00 is good and arriving at 2:01 and your day is totally ruined. That is not how anchors work. Value is on a sliding scale. We need to identify that scale. Without it we are left with either binary good/bad or an analysis of value based in a "seems like" kind of gut feel.

A common point that triggers that binary response of good or bad for a tool delivery is a program phase gate. The tool intent may have been to inform us if we have 80% confidence in the reliability goal by phase #1.

Test results don't lose all value if they are delivered a day late. They do decrease in value rapidly after that point though. But by how much and at what rate? If those test results are forecasting an epic disaster, they will still hold value to leadership discussion even at the end of the program, knowing that we can kick off a preemptive design improvement initiative ahead of the field discovering it. Yes, there will be some customers that experience the issue, but we were able to minimize the impact.

Delivery Anchor Statement

An essential element of the Delivery Anchor is that it captures the relationship between time and impact to the program or product.

Want an example of a Delivery Anchor statement? Here's a good one, followed by a bad one:

Good Delivery Anchor statement:

> The ALT test #295 will provide a statistical prediction for the percentage of units that may wear out during intended life. There are two identified high-risk wear-out failure modes in this investigation. Both will require product redesign if found to be insufficient. Redesign is expected to take an additional six weeks. ALT test #295 should be complete including one design feature optimization by 15 May 2021.

Bad Delivery Anchor statement:

> The test will tell us how long the product will last. This information is required before product release.

As the "anchored" test progresses, it is the reliability engineer's responsibility that the program and the test deliverables stay inline. The Anchoring method is great for this. It provides clear recommendations based on analysis. it will indicate if resources should be added or subtracted based on current conditions. If the test owner, let's say a reliability manager, states that "This test has a strong decay in value and should be considered for cancelation," people are going to listen.

The Value of Anchoring

How often do reliability managers suggest cancelling reliability tests? It's unheard of. But they should. It is the fastest way a reliability manager can show they are aligned with the program's needs. And to be truthful, they often aren't – same as the other roles that have had fingers pointed at them in this book. The ease of acquiring approved funding for their next test program just went way up. They have demonstrated that they can be trusted to have the program's best interest in mind. Why? They are connecting their tests to the high-level program goals with Intent and Delivery Anchoring.

Anchoring's value to a program can be summarized in this one statement:

> "Anchoring permits the team to optimize tool ROI by protecting initiatives that add great value, and flagging initiatives that are decaying in value."

Focus Rotation

The Focus Rotation method is a mechanism to ensure the Bounding goals guide the team. It also gets associated with amphibians (you'll see in a minute).

Remember the Bounding goals are the full set of program goals (features, time to market, reliability, and product cost) as set by the original product program plan. The program plan was created by sales, marketing, leadership, and R&D. The goals were carefully selected and balanced relative to each other to create a product that ensures the company grabs and holds as much market share as possible, i.e. makes money, big money.

The Focus Rotation Steps

These are the steps to the Focus Rotation method:

- List the product and program factors?
  - Cost point, reliability goal, time to market, and features are a great place to start.
- Each week one of these goals is the selected Focus Rotation goal "FRG" (yes, you can call it "Frog." This is inevitable no matter what team I work with. I've even made a mascot that we drop in meeting notes).
  - For example, this week's Frog is time to market, next week it is reliability.
- For the entire team the goal of the week is held as the #1 program priority. For each team member the Frog takes priority over their personal goal associated with their role. Yup, that's right, the project manager will have time to market as their second priority 75% of the time.

- The current Frog is present in every team meeting. This is done by discussing it briefly when the meeting opens.
- At the end of team meetings the topics discussed are reviewed in context to this week's Frog.
- Frog-related actions are assigned to individuals to ensure those topics are closed. It's preferred if the individual's role is not related to the action's topic, the Frog.

Let's take a look at Focus Rotation in action.

The team is in our Wednesday design review. The topic today is, "Why are the design prototypes going to be late? Again!"

Seventy per cent of the room is just sitting there, not a part of the discussion in any way. It's the project manager and the designers slugging it out.

They are fighting to protect the single program element they expect to be measured on. That element will be the main topic in their personal annual review. More importantly, it will be used to justify either a cost of living increase or an actual raise in their salary. This fight is personal.

Designers are measured on a design functioning successfully. Project managers are measured on time to market. This is not a debate: it's a back alley street fight.

Our fundamental issue here is that the company's complete program goal set is not represented at all in this "discussion." What's the chance the balanced goal set will be represented in the final product? You know the answer to this or you wouldn't still be reading this book.

The only way any semblance of the original goal balance is represented in the final product is if an executive, who holds this full goal set, steps in. They then have to push this full goal set onto the team. Like a parent breaking up kids in a neighborhood fighting over who gets to use the one ball in the street.

The crazy thing is that this parent is the one who told those kids an hour earlier, "Here's a ball, I want you to take turns with it, oh and by the way whoever brings the ball back to me fastest gets an ice cream." I was a team player 10 seconds ago. Now I'm going to be the one to get that ice cream.

What needs to change to align the company's objectives and the team's objectives is simple, and actually obvious. Everyone gets ice cream if you share the ball, have fun, and I don't have to stop what I am doing and come out to the street every 10 minutes.

Simply put, when the correct dynamics are in place, the team can be trusted to operate independently. Less managing and less conflict resolution for the leaders, because their team is guided by the same objectives and principles they are.

Teams also appreciate working in an environment where micromanagement isn't the norm. There have been so many studies showing that individuals operate much more productively if the external driving forces for their role come from a need to satisfy one's peers, not leadership. I don't think that dynamic is realized at all in modern organizations. Let's talk about that for a minute.

Working in Freedom and with Ownership

Individuals drive themselves harder to create results when they feel accountable to their peers and not leadership. Think about how the most resistive people in the world, teenagers, react to outside influence. We make fun of them but in fact they are a great point of reference for basic human instincts. Their hormones don't permit them to express themselves in a dosed manner It's raw expression.

Teenagers will refute every directive given by their parents, teachers, and leadership. They will just about kill themselves to gain the approval of their peers. The tribe's acceptance is the difference between life and death at a primal cortex level. If the tribe rejects you and you are outcast that is a death sentence in the wild. That's the deal for pack animals. If you aren't accepted, you're out.

The Gore Example

The influence of peer acceptance vs. pressure from leadership has been researched and experimented significantly in the business world. The results are very consistent. There is one company in particular that demonstrates this principle like no other: Gore.

You know Gore best for their Gore-Tex line of products. But they create way more than just cutting-edge fabrics. The dizzying pace at which they invent new products in multiple industries is unprecedented. They produce the best guitar strings on the planet.

They are the only brand that studied what wears out guitar strings before anything else (skin oil), and solved it. "Glide" dental floss which is proven to get food out of extremely tight tooth gaps and not break better than anything else on the market. Then there are their implantable medical devices, 40 million implanted to date. That would be 1 in 10 Americans with a Gore product implanted in their body. Filtration materials used in all industries, electronics components that are smaller, and have better corrosion and thermal management than all competitors'. This could be a five-page list, but I'll stop there.

Why are they so good at inventing and making highly reliable, advanced featured, cost-effective products so quickly?

They don't have to play by the same rules as the rest of us, and it's not fair at all. Their organizational structure is like no other. It is so unique that tens to hundreds of studies have been done on it.

Why is it so interesting? They have removed all organizational roles. There are no roles, so there are no competing goals. Serve the pack or get out.

No titles, no organizational hierarchy, no leadership-mandated program initiatives. Everything is done by the tribe and driven by peer support. How far does this go? The only role that exists is CEO, and this is just because the outside world requires it. Basically, it's there so the rest of us can interface with them, like an adapter used for highly advanced HDMI data transfer to connect with an old analog VCR.

But even the CEO selection is done based on their philosophy. The role of CEO is not filled in the traditional manner: the end of a successful climb up the corporate ladder. How could it be? There is no corporate ladder.

An individual contributor with no previous title is elected to the position. Just to be clear, someone working in a sea of peers with no title is one day voted to represent the entire company. Amazing, and this works. Internal initiatives are selected or rejected based on peer interest. If you can get critical mass in peers who want to invest time in your idea then you can progress. if people don't believe in a program, it dies. Its success or failure is attributed to the entire team. No individual wins or losses, just the team. The team is a subset of the company.

Why Don't All Companies Do This?

Here's what I guess you're thinking right now. "If this is so effective, why don't I see companies operating like this all over the world?" There is one reason. The leaders that would implement this would be required to demote themselves to individual contributors. No title, no corner office, no more big bonuses, and most of all no authority.

What would this transition be like for them? Those individuals reporting to you that you held so much power over last week? Well, today you desperately need their approval if you're going to be a contributor to a program. Even the best leaders would find that transition near impossible.

The Focus Rotation method brings one of the critical elements of the Gore ideology to traditional organizations: personal success is tied to team collaboration.

Think about Focus Rotation again. At the end of the year, you can't ask for that raise if many of your Frog actions were incomplete. You can't have successful Frog actions if the team chooses not to work with you. To get that raise or promotion, you need the team.

We're finally aligned: individual to team, team to business.

Summary

These reliability culture tools are easy to understand and easy to implement. In fact, there are likely some elements of them in your existing process. By formally introducing them to the product development program the team will be able to maintain control of the reliability initiative. Remember, this isn't the reliability initiative from the reliability team. This is the initiative set out by the business, the initiative that was identified as a key element to gaining and holding market share.

9

Guiding the Program in Motion

A program in motion, going in the right direction, with all necessary resources available isn't a sure thing. I mentioned "fire and forget" strategies earlier. They don't work in project management. The landscape we are operating in is changing. What we know about the product as it develops is changing. The market is changing.

If we make delayed large corrections to these changes, we are swerving wildly across the road. If we observe what is going on around us and make daily corrections the changes are just about invisible. This is what Guidance Bounding is about.

Guidance Bounding

We have discussed Strategy Bounding. Now we are in motion; so now, the second part, Guidance Bounding.

We're in the middle of the program and things are changing. We made our plan while sitting in the proverbial "safe, quiet, sunny harbor." Now we are out in the middle of a raging ocean with no land in sight. What do we do to stay on course?

Yes, everything is late. My budget just got robbed by those jerks in the science R&D group, and Rob has ramped up the politics game to 11 (and it's always Rob who starts that). We're also on our second unplanned major redesign. This is because no one stopped marketing from changing the specifications. Even if everything goes perfect from here on out, we are late. Anyone have Dramamine? These waves are making me sick.

The Guidance Bounding toolset will surely help here. It includes:

- Guidance Bounding return on investment (ROI)
- Program Risk Effects Analysis (PREA).

Simply put, Guidance Bounding makes sure you are headed in the intended direction and then keeps you on that path.

Reliability Culture: How Leaders Build Organizations that Create Reliable Products, First Edition.
Adam P. Bahret
© 2021 John Wiley & Sons Ltd. Published 2021 by John Wiley & Sons Ltd.

Guidance Bounding ROI

Time for a case study. This is a disguised-yet-real-life example where the Bounding ROI was applied when a design issue was found in a program, late.

The program was developing a cutting-edge electromechanical medical device. Think cyborg. (I get to work on the coolest stuff. Sometimes I can't even believe my career is real.)

This was categorized as capital equipment. The product also has one-time use components that are integrated in the larger system. Think how car tires or printer cartridges are consumed, but the central device lasts for a decade. In this case, the capital equipment is expected to last for five years. The preventative maintenance, like replacing a tire before it blows or changing an almost empty print cartridge, was in 22-month intervals.

The Plan

This medical device had a full reliability growth (RG) program planned. Twelve units would be allocated for RG in three stages. The RG program was expected to continue even past product release. Once the units were in the field there'd be a surge of data. We could grow our statistical confidence with this data at an aggressive pace. Even though the product is in the field, this confidence growth is still very valuable.

If we find the confidence not on a good trajectory we can now initiate design changes far earlier than if we had simply waited until the customer compiled the data for us. This was often delivered in stacks of angry complaints. Harshly worded letters are not the same as RG test reports, although they do hold some similar content.

The Issue

In the case of this example the RG program, it was lucky we had it in motion as early as we did. We identified a fundamental issue. The data predicted a 10% chance of the found issue occurring in the field.

Once identified, the issue was easily corrected in the design. The redesign and prototype process would take four weeks. Re-tooling and the creation of new units would be an additional 6 weeks – a 10-week total. So the program would be back on track in less than three months.

When we were in meetings to discuss what to do about this, the room did not have a lot of smiling faces. We knew we had a hard choice between correcting a known issue or going against a solid deadline for product delivery.

What do I mean by "solid"? This delivery date had some deep stakes holding it down.

What triggered our new product program was an advancement in a specific technology in another field. This occurs often.

Technology Cascade

One industry advances and its application in a second industry creates sizable benefits. This was the case with cell phones, laptops, and electric cars. It wasn't until the cell phone and laptop industries drove huge advancements in electric battery size and capacity that electric cars became truly viable.

There is a quote from GM's CEO, Bob Lutz, in 2004 stating that, "Elon Musk was ridiculous thinking he was going to make electric cars using piles of laptop batteries." Fast-forward two years and GM, Nissan, Toyota, and everyone else was frantically rushing to create a prototype electric car with those silly little lithium ion laptop batteries.

This is exactly what was happening here. Simply put, "The race was on!"

That delivery date was firm because all manufacturer's knew that was the time to beat to be the first to market. But here is what happened, just last month a competitor released their version of this new product. It was much sooner than anyone expected. And here we were proposing slowing down our program by three months. Like I said, at that meeting there weren't many smiles.

Timing is Everything

Yes, the first competing product had made it to market, but this is how that event translated to marketing's new "immovable date." What the marketing team knew from past experiences was that release in the same calendar year created a blur as to who actually released first. As time passed, people would forget which of the two companies was first if both released in, let's say, 2012. But everyone else was after 2012. Those post-2012 were just copycats who likely reverse engineered what was on the market, even though that is not the case.

More often than not, the brand considered to have created the original is also considered the best. Remember, they invented the technology.

The end of the calendar year was eight months away. Our proposed three-month delay put us over that previously achievable deadline. If we made the fix, we would be releasing next year, second place. Also known as "first loser."

But here's the other side. If we released a design with a reliability issue, it could result in our leaving that market altogether. We were sure of this because we saw that outcome with a competitor a few years back when there was a similar type of jump in the technology.

Our Choice

Be one of the first to market and have a shot at being the brand synonymous with the product: the Kleenex of tissues, the Xerox of copy machines, the Band-Aid of bandages, the Coke of colas. But. . . risk having field failures early in production that would very likely result in your losing that market entirely.

Or:

Release at a time that puts us in the middle of the pack and hope we can fight our way to the front over the coming years.

Using Bounding

Remember those Bounding tables we created? They are about to come in really handy again, because how do you make a decision like that? Everyone just arguing their opinion, with resolution coming when either someone pulls rank or there is one person left standing after the opponents tap out – single elimination cage fight anyone?

With the previously established Bounding scoring system for risk and investment, we can make a quantitative decision, our most thought-out decision. We first will adjust the scoring to represent the new landscape. Remember "closed loop control" from Chapter 8? Since the program has progressed, the rewards and their currency value have likely changed. We need to adjust them.

This 10-week investment may have had a Bounding conversion number of 5 earlier in the program. Its "cost" has gone way up now with the new information. Now losing 10 weeks means a guaranteed hard fight from midpack for years to come. A 10-week loss is now a Bounding number of 20.

The team sat down and rescored the tables to reflect "today." We then made our decisions on ROI in the same manner we did when we were Strategy Bounding.

The Results

So I bet you're curious as to what happened.

We decided to delay release and make the design change, and. . . things turned out great. Through a stronger marketing effort and leveraging that we had "the most robust technology" the company became the market leader. Marketing really came through. Their campaign chose to focus on the high investment in reliability, not letting the customer experience failures.

It was a great strategy and a "one-two knockout punch" because everyone else had both minor and major issues over the coming years. We didn't.

Program Risk Effects Analysis

We are midprogram and halfway through the Accelerated Life Test (ALT) for the wear-out mode that is our greatest concern. It is a simple Piezo force sensor that we are purchasing from a supplier, a catalogue item. This Piezo is technology that is decades old and well proven. But we have a concern, "Are we using it in a manner that matches *exactly* how it was intended to be used?" If we are planning to use it differently then all that history of proven reliability does not apply.

I am one to never forget the importance of the actual definition of reliability. It's like a pop song that gets stuck in your head. You find yourself mumbling the lyrics without even realizing it. Stand really close to me sometime when I'm not paying attention and see if you hear me quietly singing.

"The probability that an item will perform its *intended function* under *stated conditions* for a *specified interval*." I have no idea what tune that fits. Maybe "Let it Be" by the Beatles.

If any of those three qualifiers, "intended function", "stated conditions", or "specified interval" changes, all bets are off. The "probability that an item will perform" is no longer valid.

We knew we had the "intended function" correct. I was confident of that. The load range in our application was half the sensor force rating. "Specified interval" was good. The sensor was rated for four times our required life.

I was not comfortable with "stated conditions." There was a shear force in our application that was unusual. What's shear force? It's a sideways force that can split things into layers.

The ALT test would be extremely valuable in predicting if this shear failure could happen in our design. Unfortunately, the ALT test turned out to be extremely valuable. It gave us an early heads-up that we were in trouble with this design. When we root caused the issue in the test, it was clear "cards" were sliding past each other when under a shear load.

What Now?

So here we are at the executive steering meeting all staring at each other with this new information. The question that was just left hanging in the air? "What do we do?" It became very clear we came with problems, and no solutions. Not smart.

Obviously, we're going to do a redesign and run it through ALT test again. The big questions, "Do we run it through a full ALT test or just far enough to know it is better than before?" and, "How much program time are we willing to lose?"

ALT testing is expensive and takes a long time. In Bounding number language, "It's a big B#i (Investment Bounding Number)".

So everyone is dug in and starts campaigning for what served them best. Marketing wants it out the door, engineering wants it to be proven to be durable, Sam is insisting that all meetings have complimentary cookies. Why is he here? Sam works at the front desk.

This is the argument the project manager made, "We release on time, and start testing now. If there is an issue with the new design, we will know ahead of the field. We can then get a fix out there fast with minimal field damage." Sound familiar from my last example? This is the counterargument reliability is regularly debating against.

Just Let It Go

My response, "Ummm, so basically you are saying plan for a recall?" It was clear that no one was arguing for what was actually best for the business. Especially Sam. He was hell bent on wrecking the snacks budget in Q1. Why is he still here? Someone throw a cookie in the hall and get rid of him.

So clearly, I didn't have faith that these types of issues would be resolved with the best interest of the company in mind. It will just be another team slugging it out until a victor is declared, yours truly included. I can openly admit that I got in there and swung as hard as anyone else once things got going. Even if you are not an advocate of violence, when a bar brawl breaks out you have few options.

After going through this for the 43rd time in my career I wanted to find some kind of a solution for this predictable scenario. I mean, seriously, when do you expect to test something and find out it is perfect in the first round? Almost all programs only allocate enough time for a single round of testing. There's never a mechanism for dealing with the bad results.

So I thought of an idea. I would like to say it was a spark of genius, but like most good ideas it is a natural progression of things everyone knows. I labeled the technique PREA (Program Risk Effects Analysis). The problem this technique solves is, "Why don't we access the levels of risk to the program and product in advance of big program decisions?" If we had this we could then use it to quantitatively balance how much risk to the program vs the product we are willing to accept.

Fully Access Risk

A PREA looks to fully access this risk. The difference is that it is program centric. I'll repeat that since it is a critical differentiator from other risk tools, *A PREA accesses risk to the program's success not the product's success*. The PREA objective is to assist with deciding what to do when the product design and program goals end up head to head pushing against each other.

The PREA method is based on ranked factors methodology, like FMEA. A common question that leads us to use PREA is, "Do we or do we not add resources to address the issue in front of us? Designs want resources; programs don't want to give it up. Those are the laws of nature. This question typically leads us to evaluating the impacts to warranty cost, sales, time to market, and the product doing what it is supposed to do in the customer's hands.

Unlike the Design Failure Mode Effects Analysis (DFMEA), we don't know what the situation is that we are planning to analyze in advance. When a DFMEA is occurring, we have a basic design or concept in front of us. With PREA we are left with the first big challenge, "How do I rank risk factors for a conflict that hasn't happened yet?"

What I realized when looking a little closer to how DFMEAs work is that they have two distinct phases. This isn't always clear, because the first phase is quick and the second phase is very long. They seem to just blend into one. When a DFMEA starts, it is always the first task to decide what the ranking definitions are. What are our "severity," "detectability," and "occurrence" scales? This is our Phase 1. In a PREA this is a much more significant step. It has to hold up for future unknown analysis topics.

For the PREA this first phase is done completely separately from the second phase. It is actually the one that is longer. With these scales complete the foundation for the PREA process is set and ready to be called upon when a program conflict arises. The program, not the design, has a living breathing FMEA type of analysis going alongside it at all times. It's quiet but ready to engage when needed.

The key was going to be making sure that we upheld the DFMEA rules with the creation of these scales. They also have to be done completely in advance of the program beginning. The guidelines would be the same as the DFMEA. A cross-functional team will participate in the scale creation. Keep in mind that anyone who you believe will have an interest in negotiating during program vs design issues should be a part of this phase. It needs to be evident that how the scale resolves the issue is based on everyone's input. There cannot be debate about the scales when they are in use in phase 2. That defeats the entire purpose.

Here is a second factor that makes this tool so important. It addresses an open secret that exists in all programs in all industries.

Program Freezes Don't Work

These are "design freezes," "concept freezes," and "manufacturing prototype freezes." If the sensor scenario program conflict I described earlier is a bar brawl, "freezes" are then full-blown, pay-per-view boxing matches. They are both "conflict," but the "freeze" is better organized, scheduled for a specific date, and the competitors train for it. If you have worked in programs littered with "freezes" then you know none of that is an exaggeration.

Why is it an open secret? Because leaders include freezes in programs with the hope of "policing" the process. This has never ever happened. Everyone knows what they actually are. Freezes are the bell rung to let everyone know the fight has begun. Can we just call freezes fight bells, so we know we are all on the same page?

Or. . . we could actually make them work by adapting how we use them.

What is the purpose of a freeze gate? In a program, freeze gates are intended to be hard stops for a type of work. A design freeze is the date all design work is completed by, for example. In reality I would say a freeze is at best a suggestion, like a stop sign in a cattle drive.

Freezes are the dates we begin the negotiations for more time – remember "fight bell." When the freeze comes up everyone is ready to start campaigning as to why they should have an extension. In fact, they have been working on their case for weeks. The program leaders are left with hunkering down behind piles of sandbags that they also have been working on for weeks.

In all this fuss, the big loss was opportunities to make program corrections when it can change outcomes. What if these negotiations were started earlier? Yes, there is still a gate called "freeze," but we can adapt to it if we don't wait until it is upon us before discussing the conflicts between design and the program.

How can we do this? How about a phase that addresses that these changes require more sacrifice as we get closer to the gate? How about a "chill phase"?

The Chill Phase

The chill phase is the period of time leading up to the "freeze." The chill phase more formally acknowledges that proposed changes require greater benefit to be validated as we get closer to the freeze.

Realizing how easy it is to bring that into the light makes it seem ridiculous we let this go on so long. Let's take that ad hoc process of whispering discussions among the team until the freeze and make it a planned organized discussion based around an agreed upon method.

Like I said in the beginning, "I haven't invented anything great here. I'm just applying known tools to new situations."

How does it work? In simple terms, "You will need a stronger argument to make a requested change in resources or schedule as you approach the freeze date."

But how do we make it quantitative so it's not just re-labeling the same bar brawl as a "club debate"?

This is how. If the engineering team wants to change the product one month before the design freeze, it had better be because you think the product is going to kill a customer. And you can prove that based on our previously agreed-upon scale. If they are requesting resources and time for that same change six months before the design freeze, less justification is needed.

So I experimented with creating different types of scales that could be used to capture the critical factors for the chill phase. Some got complex. When I see my attempt to capture a concept in math or even in dialogue start to get complex, I scrap it. That complexity is coming from me trying to explain something I haven't figured out.

What I found is that if we create four factors, a simple "go, no go" balance equation can be made. If one side of the equation is larger than the other, that side wins. If one side of the equation represents benefit to the product and the other benefit to the program, we have a quantitative way to decide which way to go.

A simpler way to set up this comparative analysis is to simply subtract the two sides and have that equate to a variable. If that variable is positive then the first element was greater. If it is negative, then the second element was greater. That variable is called the "balance factor."

PREA Tables and Calculations

This is what the tables and equations for the PREA factors look like (Figures 9.1 and 9.2):

Each factor has three levels based on the predetermined PREA tables. The factors are "warranty failure expense," "sales from reliability," "time to market impact to sales," "features impact to sales." The warranty and sales are impact to the product. The time to market and features are the impact to the program. The two factors for each are multiplied together. Then, the two resulting terms are subtracted from each other. As mentioned, the resulting number is the "balance factor." If the balance factor is negative then the freeze should be held and the requested changes to the program are rejected. If the number is positive, the requested program changes should be executed.

Let's discuss making the tables and scales.

Each table represents a factor. The product has "impact to warranty" and "impact to sales" as its two factors. The scale levels for each table will be 1, 3, and 5. The two factors that will represent the "product reliability" will be "impact to warranty" and "impact to sales."

Table Creation

Warranty will have "low warranty" associated with a value of one. "Low warranty" represents an annual warranty cost that is 25% below what was predicted, a nice surprise! Determining this value can of course be difficult. A process I would use in this specific case is interviewing individuals in the quality department. These are the individuals closest to the necessary data. We could ask, "In a past program, 'What was one of the best actual vs projected warranty expense ratios?' "

It is valuable to include leadership when creating these scales. Leaders know a lot about program loss. They've surprised me by saying, "It would be a first if actual warranty matched projected warranty." That's a red flag. Why do they lack self-awareness? This table creation exercise is going to have some big peripheral benefits regarding mindset.

A word of caution: look out for the eternal optimist. These individuals who say, "Everything is great" are the ones knitting the wool that seems to be covering everyone's eyes. The Yes man. They exist in every organization, because as humans we always want to hear that everything is going to be alright. They exist out of necessity.

A score of "3" (Moderate) is a 10% over our warranty projection. A score of "5" is 30% over our budget (Figure 9.3).

Developing the sales scoring table can get a bit heated. It's hard to associate reliability issues to actual lost sales. For this one, I just say include as many opinions as you can.

Impact to PRODUCT RELIABILITY

Warranty			
Scale	1	3	5
Description	Low Warranty Cost and highly reliable product "Bullet Proof"	Low sale loss due to warranty. Some of poor reliability	Loss of future sale and high dollar warranty cost

Sales			
Scale	1	3	5
Description	Target market share	30% below target market share	50% below target market share

Impact to PROGRAM GOALS

Features			
Scale	1	3	5
Description	All Planned Features	Reduced functionality features	Missing key Features

Time to Market			
Scale	1	3	5
Description	On-Time	6 month delay	1 year + delay

Figure 9.1 PREA tables.

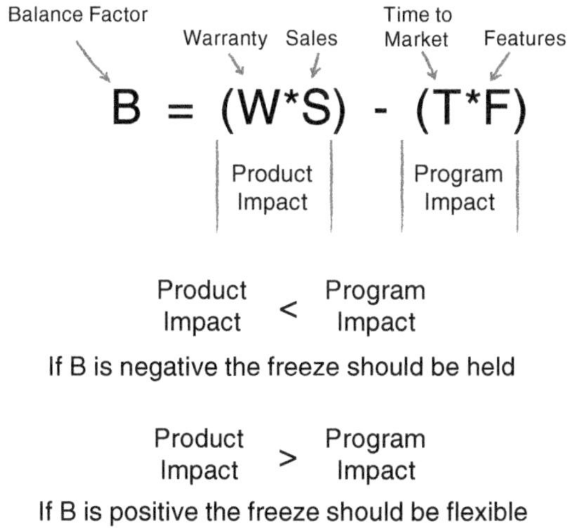

Balance Factor
Warranty Sales
Time to Market Features

$$B = (W*S) - (T*F)$$

Product Impact
Program Impact

Product Impact < Program Impact

If B is negative the freeze should be held

Product Impact > Program Impact

If B is positive the freeze should be flexible

Figure 9.2 PREA balance equation.

Warranty			
Scale	1	3	5
Description	First year hits warranty budget	First year is 10% over warranty budget	First year is 30% over warranty budget

Figure 9.3 Warranty table.

Sales			
Scale	1	3	5
Description	Target market share	10% below target market share due to reliability issues	25% below target market share due to reliability issues

Figure 9.4 Sales table.

Features			
Scale	1	3	5
Description	All Planned Features	All features but some reduced functionality	Missing a planned feature

Figure 9.5 Features table.

Time to Market			
Scale	1	3	5
Description	On-Time	6 month delay	1 year + delay

Figure 9.6 Time to market table.

We can create a scale where 1 is everything went well and there is no clear link between reliability performance and loss of sales. A score of 3 may be that we can associate 10% loss of sales due to reliability issues. This association might be based on online feedback and how well that is associated to customer's choices of brand when they buy again. A score of 5 will be associated with a 25% sales loss due to reliability (Figure 9.4).

For the "program goals" tables we will have "features" and "time to market." The scale will range from a released product that includes all planned features (one), to one that is missing one or more key features entirely (5). A 3 includes the planned features but without some convenience (Figure 9.5). An example could be an air conditioner remote control, the new feature, but without a temperature readback on the remote.

Time to market will be a scale of calendar time that is rooted in historical examples. One may be "on time, three is 'six-month delay,' " and five "one-year delay." There will be project managers happy to share war stories about programs that almost got cancelled, battles between features and schedule (Figure 9.6).

Evaluation

With these tables in place we were able to evaluate the impact of a full second round of sensor ALT testing.

What were the product and program factors for the Piezo sensor life issue?

On the product side: warranty 3. This value came about based on a series of data supported discussions with the engineers.

Even with the few life test data points we had a distribution for premature wear-out could be made. The distribution was a lognormal curve with a long left tail. Placing a known good life curve showing a normal distribution, we were able to estimate a percentage of premature wear-out for the first year. These failures would occur if the worst units were in the hands of the highest cycle users (Figure 9.7).

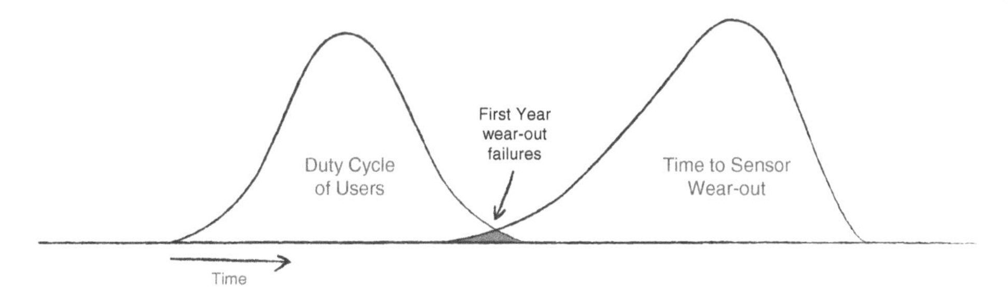

Figure 9.7 Wear-out percentage.

Layered Piezo Sensor Life PREA Tables

Impact to PRODUCT RELIABILITY

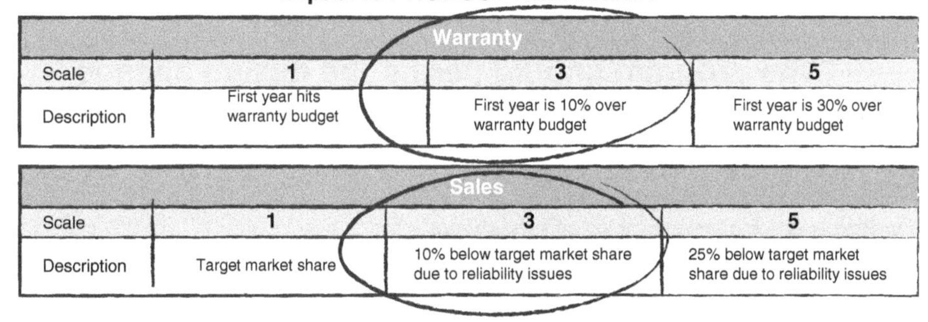

Warranty			
Scale	1	3	5
Description	First year hits warranty budget	First year is 10% over warranty budget	First year is 30% over warranty budget

Sales			
Scale	1	3	5
Description	Target market share	10% below target market share due to reliability issues	25% below target market share due to reliability issues

Impact to PROGRAM GOALS

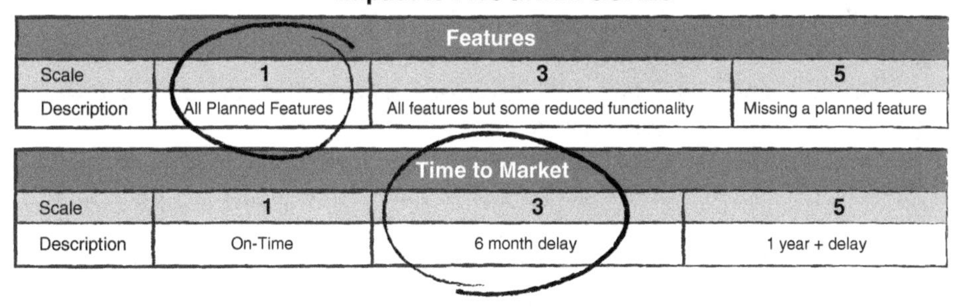

Features			
Scale	1	3	5
Description	All Planned Features	All features but some reduced functionality	Missing a planned feature

Time to Market			
Scale	1	3	5
Description	On-Time	6 month delay	1 year + delay

Figure 9.8 Piezo tables.

We estimate this dark area will put us 10% over our warranty budget, just from our one found failure mode. Remember there could be others we haven't identified yet.

Our sales factor was selected to be one. We didn't believe that this issue would affect sales. It will take some time to become evident.

Features were also selected to be a 1. Through discussion with the team, it became clear that resources would not be pulled from any other part of this product line's development to address this issue. The product's design had already been frozen.

What were the selected values from the tables (Figure 9.8)?

On the product side: warranty 3, and Sales 3

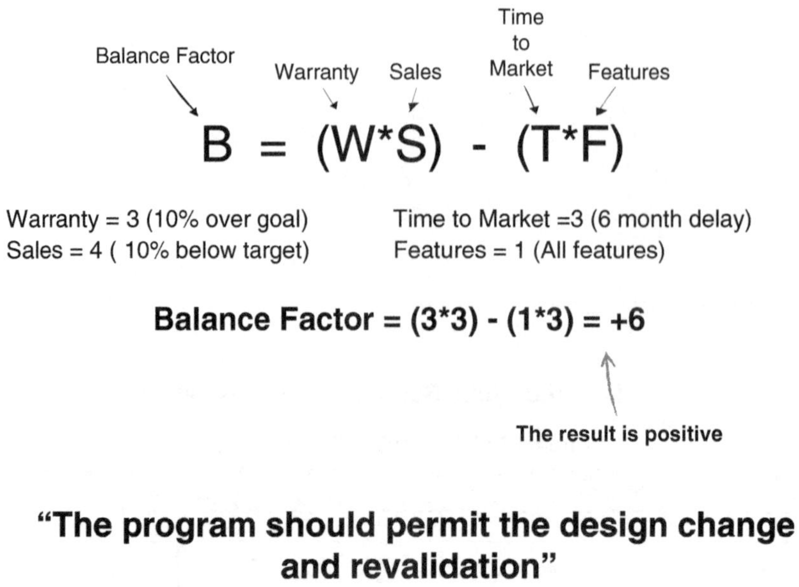

Balance Factor → B, Warranty → W, Sales → S, Time to Market → T, Features → F

$$B = (W*S) - (T*F)$$

Warranty = 3 (10% over goal)
Sales = 4 (10% below target)

Time to Market =3 (6 month delay)
Features = 1 (All features)

Balance Factor = (3*3) - (1*3) = +6

The result is positive

"The program should permit the design change and revalidation"

Figure 9.9 Piezo balance equation.

On the program side: feature 1 and time to market 3

Warranty: to date all have demonstrated a capability to achieve use life and the random failure rate reliability goals. So warranty for the first year will only be minimally affected.

Sales: the issue would have a relatively minimal impact to sales. The inclusion of the full functionality of the product would drive great market share. It would be sales in following years that may be hurt.

Features: if the life issue cannot be resolved either the program will have to go back to the R&D stage or the product will have to be advertised as having a shorter use life. It was not an option to remove this essential feature.

Time to market: it is estimated that the redesign and retest process would take approximately six months. Six months was chosen because the R&D team felt they understood the root cause issue and had a solution they were confident in.

These values in the balance factor equation result in a value of positive 6 (Figure 9.9).

We should proceed with the redesign and full testing to ensure the product does not have a premature failure mode. This is what is best for product, program, team, leadership, and the business.

Summary

Fire and forget worked when we used cannons in war. Actually, they didn't really work that well: we just didn't have any other option. As soon as we figured out how to track the progress of projectiles, vehicles, and program initiatives, we looked for ways to have those progress reports steer the ongoing event.

So ask yourself: are you firing any cannonballs in your programs that could be guided rockets?

10

Risk Analysis Guided Project Management

We have limited resources and limited time. We would like to select and adhere to a method to choose where to apply resources. I firmly believe that assessed risk is the best way to do this. Using risk assessment to guide program activities is a process of:

- Selecting tools that can translate product performance and program success risk to quantitative values.
- Use these values to prioritize the most critical areas of the product and program to address.
- Create a summary of these quantitative risk rankings that assigns specific tools to measure and mitigate these risks.
- Create a program plan that applies the available resources to these items in rank order until time and resources run out.

Failure Mode Effects Analysis Methodology

A Failure Mode Effects Analysis (FMEA) looks to take speculative failure modes and their effects and make them quantifiable so that they can be ranked from highest to lowest risk. FMEAs can be used in programs for many purposes.

I have noticed in my career that simply uttering the word FMEA creates a "cringe" in the room. I would believe that if I were to choose the phrase that best describes how many teams feel about FMEAs it would be what was said to me after I did a "lunch and learn." "I get what you are saying about the value of FMEAs, but we have been a bit traumatized by them in the past." Traumatized?! The team lead actually used the word "traumatized."

But that is not an uncommon experience and the negative effects are real. FMEAs are a true double-edged sword. I sincerely believe they are one of the most valuable cornerstones to any reliability and product development program, if done correctly. Like anything else that is powerful, it's just as easy to cut off your own leg with it if you don't know what you are doing.

Reliability Culture: How Leaders Build Organizations that Create Reliable Products, First Edition.
Adam P. Bahret
© 2021 John Wiley & Sons Ltd. Published 2021 by John Wiley & Sons Ltd.

Simply put, a badly run FMEA is something you would wish upon your competitors. It wastes time, drains morale, and doesn't provide any valuable program or design input. What more could you hope for for the other guy with a product next to yours on the store shelf?

There are many flavors of FMEA. The three most common are Design (DFMEA), Process (PFMEA), and Use (UFMEA).

Design Failure Mode Effects Analysis

Risk analysis tools like Design Failure Mode Effects Analysis (DFMEA) are fundamental to understanding the greatest threats to the success of the product in the field. By identifying risk, resources can be allocated in a way that matches the need. Just assessing risk isn't valuable to an engineer. Everything has risk. Which items should I act on? What level of risk is too risky? A tool like a DFMEA can be of significant value to the team, as long as it is recategorized as being primarily a project management tool.

The value to the engineer is that the entire multidisciplinary team is providing input (read "ownership") into the risks in the design. By doing this there is an automatic resource, time, and money allocation to specific design parameters. Actions by the entire team associated with assessing and reducing this risk will be agreed upon.

In no way did the exercise mitigate the risk. That's the designer's job. The designer wants the DFMEA so they have input, buy-in, and do not have to campaign for time or money to spend "more" on developing the product. It is a critical tool in the designer's process. With a DFMEA-based risk analysis, a designer can ensure they are working on the right things at the right time.

We need to constantly remind ourselves and our team what the intent of a DFMEA is during the process. It is simply a project management tool. The purpose of a DFMEA is to sort technical design areas in order of greatest risk to lowest risk.

That's it. Using it to solve engineering problems is a bad idea. Using it to simply document that the team has thought of every possible feature as preparation for a yet-to-occur investigation is also a bad idea.

The guidelines to performing a DFMEA well are:

Have an Experienced Facilitator Who Is Only Facilitating

They should not just be one of the team members that has volunteered (or been volunteered). This should be an individual who has taken DFMEA facilitation courses. I can't express how valuable this is to an effective DFMEA. The return on investment (ROI) for having a trained facilitator is easily 1000/1. Having a DFMEA go on twice as long as it has to is a lot of wasted working hours. This does not include broken morale and a less-actionable DFMEA.

The Facilitator Should Not Be the Scribe or "Spreadsheet Master"

This should be a second individual who is not a part of the DFMEA review team. It is almost impossible to effectively facilitate if you are also managing documentation. Trust me, I have tried and I have failed.

Don't Let Conversations Go So Deep that 90% of the Room Is Just Listening Without Being Able to Contribute

These conversations are like handing out sleeping pills. After about three of them have occurred you are lucky if people haven't found reasons to leave the room entirely (OD). If a technical debate is occurring between only two or three people, it should stop. It can be picked up in a private offline session.

They can invite others to attend the independent discussion if they like, maybe even people not in the DFMEA. Then bring the resolution back to the next meeting to be discussed and included. By doing this consistently the DFMEA will be cut to one-third the length of traditional, not correctly facilitated, DFMEAs. The engagement level will stay high for the entire team throughout all sessions.

FMEAs are significant time commitments and can become tedious, even when done perfectly. They tend to have an upward spiraling energy or a downward spiraling energy. If the trend begins to go down, the effectiveness quickly diminishes. Seeing a reduced value to the activity only further encourages team members to disengage. Next thing the team is just feeling tortured to be there and it is pulling teeth to get any participation. Once that scenario has occurred recovery is very unlikely.

If the team is having their opinions heard and they feel they can contribute on each section, they will maintain an excitement to stay in the game and have their input heard. This discussion and healthy debate will pull people in, sometimes even pulling in people who weren't part of the original team but whose expertise is needed to resolve a debate. It can be really exciting and fun.

Make a Scoring System that Is Meaningful, Not Standardized

The three base scoring categories when ranking DFMEA line items are "severity," "likelihood of occurrence," and "detectability." A scoring range may be selected that is 1–5 or 1–10. The range has to be consistent across all three categories. In advance of the DFMEA, definitions should be created and agreed upon by the entire team. It is a horrible quagmire to be debating ranking definitions mid-DFMEA. What will occur is a need to go back and reevaluate all previous line items per the changed definitions. It is a huge waste of time and another easy way to sink morale.

The definitions don't just need agreement from the team. They have to be relatable to speculative situations. We need examples for each ranking situation to be able to calibrate what the correct score is. If a 3 for severity means, customer can't complete task and customer service must be called. That is something you can relate a specific failure mode outcome to. If the definition is "Customer complaint," that could mean anything. Are they complaining to their coworkers? Do they file a complaint only after it happens three times? Does a supplier representative help them solve it on the phone or is a service technician dispatched? With a definition that vague, a scoring of 3 for severity will end up having completely different meanings for separate line times.

The Scoring Is Comparative, Not Absolute

The value of the scoring system is that the risks can be prioritized relative to each other. I have seen many teams get tied up because they want the DFMEA scoring to be translatable to other hazard or risk analysis being completed in the program. Many of those hazard analyses are required per industry specifications and have previously fixed definitions so risk, hazard, and impact can be compared to industry standards.

If we insist that the DFMEA risk definitions align with these other definitions, we lose the flexibility to create good definitions. Good definitions best support creating a resolution in ranking that permits a good sorting of risk to the program and design. I highly encourage permitting the DFMEA definitions to stand independently as much as possible of any regulatory rankings.

Reliability Design Risk Summary

Because I am a fan of the FMEA methodology for risk analysis, I eventually created a custom version to aid in the development of reliability programs. I originally would call them "reliability focused DFMEAs." The idea was to focus on risk that can be categorized as reliability and would likely have a tool from the reliability tool kit in its actions.

In a regular DFMEA or PFMEA, many of the actions use the reliability toolkit. But if our intention is to create a foundation for a product reliability program, and that was all, we could be much more efficient.

The technique evolved into Reliability Design Risk Summary (RDRS). RDRS is effectively a reliability focused DFMEA. I could have just called it "DFMEA-lite" but I found two issues with that. The first is that anything said after FMEA is usually mentally blocked out just as a reflex. The second is that there is always a regulatory/documentative accountability for an FMEA.

FMEAs are often used to create documentation that the team has "thoroughly assessed reasonable risk." So FMEAs are created with a feeling of future regulatory investigators listening in on the conversations in the room. Evidence of this is that no team has ever allowed the term "explode" or "catch fire" to be used as an effect of a failure. What are usually used are benign terms like "thermal event" or "spontaneous self-disassembly."

This is not the language of teams that are openly communicating. These are defendants in a proverbial courtroom. You can see how disassociating from the term FMEA has its benefits if our goal is to truly assess where the greatest risks to product reliability performance is.

The Objective of RDRS

The objective of RDRS is simple and concise. Collect and rank the team's perspective on greatest risk to product reliability performance in the customer's hands. To do this and stay on task it is important to reiterate the purpose of reliability testing and analysis. "Reliability is the study and control of variabilities that affect product performance."

In an RDRS we are looking to identify what actions we should take to study and control the variabilities and their effects on the product's performance. The structure looks similar to the DFMEA with part, function, failure, cause, effect; and scoring for severity, detectability, and occurrence. The big difference is that we only outline risk concerns associated with the variabilities of use, environment, and manufacturing.

Each line item must have an action that can be addressed with a reliability tool. When the RDRS is complete, there will be a very specific prescription of what to do with your reliability time and budget. To ensure this information is understood outside the group, an RDRS summary report will be created. They are so brief that I even call them "memos" most of the time. Again, I'm looking for efficiency. If a memo will do that job then that's what you are getting.

Three Ranking Factors

Three ranking factors are used, similar to a DFMEA. A DFMEA uses "severity," "likelihood of occurrence," and "detectability." The RDRS uses severity and three variants on "occurrence" It will capture likelihood of occurrence due to variability in (i) manufacturing, (ii) use, and (iii) environment.

We can't forget that we are evaluating how variability of use can affect performance. This means we have to talk about what other uses a product will be put to beyond what it was originally intended for.

We want to speculate how the customer may use the product in unexpected ways. I intentionally didn't use the word "unusual." It may actually be a common way to use the product. We just didn't expect it. It's on us for not considering it. Run through the scenarios of use and permit the team to be creative. The third input for variability we will evaluate is environment. This again will quickly lead to the imagining what kind of extreme environmental conditions the product could be exposed to.

Maybe it is an indoor appliance. What happens when it is used on a porch at a beach house? Humidity and salt are killers and not generally found in a normal kitchen environment. It should also be considered that the percentage of cases with this condition may not be as small as suspected. Could this product be great for a kitchen in a boat (galley)? Before we know it, we will be taking this "unexpected" environmental case and finding that maybe it should be a part of the design requirements.

Scoring and Evaluation

The process continues by parsing out the system and evaluating each line item by these input variabilities. When the full system has been evaluated and the scoring is complete, review the full document with the team. It is an important step to look at the scoring now in a comparative manner to the other line items/subsystems. The value of the scoring is not absolute. In all honesty you could erase the scoring tables after the analysis is done.

They could be erased because it is the comparative value of the scores that has value. Once we have the ranked items in descending order we have the information we need. We then apply resources and a schedule to the items, top to bottom, until we run out of people or time. The numbers are no longer important after the items are in order.

Risk Priority Numbers

For this reason it is very important to go back and evaluate that the risk priority number (RPN) scores make sense when being compared to their neighbors. If an item seems too low or too high on the RPN ranking, reevaluate it with the team. The definitions in the scoring table are there to bring consistency to the scoring. Keep in mind that the interpretation of what they mean may change as the process is executed because specific instances have been associated with specific scores.

HALT Testing

The results from the analysis are ready to direct what the most critical reliability risks are to the product. We want to have a new column that lists what reliability tools will be used to address the risk. Highly Accelerated Life Testing (HALT) may be used if it is found that electrical connectors coming loose is a big concern.

We could construct a HALT test that uses incremental vibration to identify connectors that come loose at each level of energy input. With the connectors ranked by robustness relative to each other, the team can look for opportunities to improve the ability of the connectors to stay in place under stress. The solution may be to simply tie the wiring harness down in a different manner so the free mass can't pull on the connector as easily.

The Benefits of RDRS

One of the most valuable aspects of this method is that there is now a connection between the reliability tool and the need to mitigate a specific risk. When investment in a tool comes into question, it doesn't need a passionate defense from a red-faced reliability engineer. The RDRS can simply be brought up and the risk the team has assigned through evaluation can be shown.

The question is then, "Is the investment in this tool worth mitigating exposure to this risk?" A very important word in that previous sentence is "team." The RDRS is the summary of input from the entire team. To dispute the ranking is to argue against the entire team that participated. It's not an argument against an individual.

Process Failure Mode Effects Analysis

A Process Failure Mode Effects Analysis (PFMEA) evaluates how variability in the product manufacturing process can affect the product's performance in the field. It is conducted through the quality planning phase as an aid during production. The purpose is to analyze and correct the possible failure modes in the manufacturing process, including limitations in equipment, tooling, gauges, operator training, or potential sources of error.

The format of a PFMEA is very similar to a DFMEA. A PFMEA uses the format of identifying quantitative values for categories such as "severity," "occurrence," and "detectability." What is critical to abide by is what the analysis is looking to include in its scope. Because of this similarity between DFMEA and PFMEA, it is very easy for a DFMEA to include elements of a PFMEA, and vice versa. Some of the most heated moments I have experienced in a DFMEA is arguments regarding whether we were now venturing into process for our root causes.

Use Failure Mode Effects Analysis

A Use Failure Mode Effects Analysis (UFMEA) evaluates how variability in the use of the product can affect performance. If you recall, use variability is one of the big three variabilities (i.e. manufacturing, use, environment) that reliability looks to study the effects of. When reliability statements are made, it is important to reference use case. But it is important to remember that use cases are how the developers envision the product being used, not how it may actually be used. The UFMEA ensures that time is taken to speculate about all the possible applications, uses, and abuses the product will experience.

I believe that some of the unusual product uses that are shared in the "what were they thinking" conversation are:

- Using a lawn mower to trim hedges
- Drying an item with a microwave
- Using a makeup brush to clean in-between computer keyboard keys, thus introducing makeup powder into their keyboards.
- Using a hairdryer to custom fit your plastic frame glasses. Hairdryers can also be used to remove dust from items. Can dust be flammable? Is the hairdryer company accountable for the house fire caused by heating dust on books in a bookshelf?

And most of these were before we all became fascinated by the idea of #lifehacks. Which basically is just a rather risky trend of misusing products. I've never been on or contributed to a life hacks forum.

Failure Reporting and Corrective Action System

A robust Failure Reporting and Corrective Action System (FRACAS) system will ensure the organization knows who should be working on what. FRACAS was created because without it field issues will be reported back to the teams through multiple paths with varying degrees of detail and to a dynamic audience. This well-intended but chaotic system is driven by individuals trying to get the right information to the people best placed to address specific issues, and yet has some unwanted side effects, such as:

- Critical information being lost.
- Incomplete datasets.
- Relevant reports/data not getting to the correct audience.
- Duplicate initiatives.
- Field failures being in a "pull only" system. Individuals have to engage with the field to find out what is going on.

FRACAS looks to address all these issues by applying a systematic process. The fundamental components of a FRACAS system are:

- Capturing critical information about failures.
- "Scrubbing" the data. This is a process of filtering the information, grouping and prioritizing it so it can be considered valid input for analysis.
- Connecting the data to the appropriate recipient.
- Maintaining a connection between new data and continuing investigations.
- Securely storing information associated to observations, data, analysis, reports, summaries, and corrective actions.
- Ease of access. The FRACAS system must make it easy for any team member to access a dataset or investigation of interest.

Modern FRACS systems are often Web-based and can be accessed by the entire team. They tend to have functions that go beyond the basic FRACAS components. Often project management, quality assurance, and R&D functions are included. This creates a comprehensive system that captures an issue from first incident to validation of the final corrective action and beyond to improvement of "best practices."

Root Cause Analysis

The *Titanic* had two sister ships, the *Britannic* and the *Olympic*. A woman called Violet Jessop, a nurse and a cruise liner stewardess, worked on all three.

- The *Olympic* crashed into a warship whilst leaving harbor but was able to make it back.
- She was on the *Titanic* as it sank and is referenced in the *Titanic* film, a stewardess that was told to set an example to the non-English speaking passengers as the ship sank. She looked after a baby on lifeboat 16 until being rescued by the *Carpathia* the following day.
- When the *Britannic* hit a mine, triggering the "abandon ship" command to be given, the lifeboats hit the water too early. As the ship sank, the rear listed up and a number of the lifeboats were sucked into the propellers. Violet had to jump out of the lifeboat she was in and sustained a serious head injury, but survived.

She was on board for all three incidents in the space of five years.

So it's pretty safe to assume that Violet, a factor in all three events, is the cause of ships sinking. Right? It's hard to argue the odds of that "factor" in all three incidents being random. So since we are in a rush to improve ship safety we are banning the carrying of passengers or crew with the name "Violet."

Reaching a Wrong Conclusion

Sounds like a foolish conclusion and mitigative action. But... I imagine a lot of you reading this have seen the desperate search for the root cause of an issue lead to a premature conclusion and corrective action. It goes something like this.

1) Chronic issue in design/prototypes/manufactured units.
2) After weeks of desperate investigation and weekly team meetings that include multiple levels of the organization, someone finally finds what looks like a common factor to all found errors.
3) It's mentioned as an "idea," "thought," "possibility," "observation," in an update meeting.
4) "Must be it! We don't have much more time left to solve this. Do what it takes to correct it." The odds of this being present in all cases and not being the issue are unlikely at best. "Sure do some testing in parallel to implementing the solution to prove we are right."
5) Action is taken, the issue continues to appear in product, and mysteriously no one remembers being in the group that gave the "go-ahead."

Reaching the Right Conclusion

How do we protect against "action" on incomplete information? Root Cause Analysis (RCA) is a process for identifying root causes of known problems or events. The intent is ultimately be able to permanently prevent the failure/issue from occurring again with knowledge of the root cause(s). The goals for RCA is to analyze the problems or events to identify:

- "What happened?"
- "How did it happen?"
- "Why did it happen?"
- Actions for preventing reoccurrences.

First, we need to create classifications for information. Without labels anything that is "hopeful" quickly gets promoted to a "proven cause and effect relationship." This usually takes about one meeting cycle. If specific terms are used for information with different levels of credibility/assuredness then this "promotion" affect won't happen. It is important for the "information" status changes to be reported and documented. All accountable team members need to have a say on status change.

Some suggested definitions:

- **Observation:** an individual performing the investigation or data analysis has noted something of interest.
- **Hypothesis:** a statement of cause and effect for an "input" and "response" that is yet to be proven.
- **Confirmed hypothesis:** a hypothesis that has a statistical "p" value based on a completed "designed" test. (p value indicates likelihood of results being random.)
- **Validated relationship:** a statistically validated relationship between inputs and responses through experimentation.

The Stages of RCA

The RCA process consists of several stages.

1) First identify the problem. It is critical at this stage to not include speculation or opinion. We are simply looking for the facts based on observation or measurement.
2) Once the problem has been identified, information (evidence) will be gathered. This information may be recorded data, observations, or reports.
3) With a comprehensive information set, the RCA team will identify all issues that contributed to the problem. These "issues" are the building blocks of the actual failure or incident. Throughout history, catastrophes are rarely attributed to a single issue. There is almost always a series of mistakes or issues that lead to the final outcome. This is the case as well with even the simplest of failures. No one has ever tripped and fallen with a single root cause being the reason. It takes having an obstacle and a distraction at minimum to make the event happen. Get them all out on paper and begin to put them together in a logical timeline or sequence.
4) There is now enough information to create a hypothesis on root causes of the issue. Experiments will then be designed to confirm if this hypothesis has a statistically significant relationship to the failure.
5) The statistically significant relationship will then be validated through further testing and analysis.

6) Proposed solutions will be tested to verify that they mitigate the issue through addressing the root cause.
7) Then it's implementation of the solution(s) and monitoring if the problem has been fully and permanently addressed.

RCA is not a single method but, in fact, a group of methods. The most commonly used are (in order of my favorites):

1) **5 Why Analysis:** the reason this is my number one is because it is an element of the fundamental human investigation technique. How do I know this? Because if you have ever spent more than 30 minutes of time with a kid between the ages of 3 and 9 you have been on the receiving end of 5 Why Analysis in action. I used to do this so much with my two very inquisitive scientists that I turned it into a game to see if I could get all of their inquisitions back to the "Big Bang." We did this so often that they would actually cheer me on in hopes that I could.
2) **Change Analysis:** this is my second because it is also based on a very fundamental principle: "What changed?" If a problem previously didn't exist and now it does, it is very likely that something changed. There are many issues and factors that will be collected in the initial RCA phases. Bringing the ones that have changed to the top of the list is a good way to increase the chances of finding out which ones are significant earlier.
3) **FMEA:** I cover FMEA and its derivates extensively in Chapter 9. It is an amazing tool for thoroughly capturing observations and speculations about the causes, failure types, effects, and likelihood of occurrence. Most importantly, it ensures that this process is done with a diverse cross-discipline team that has many different vantage points to the problem and its effects.
4) **Fault Tree Analysis** (**FTA**)**:** this is often referred to as the "counterpart of an FMEA." An FMEA looks to identify all the possible ways a system can fail. An FTA takes a singular type of failure and then builds a logic tree that identifies all the ways, sequence of events, that could lead to that single failure.
5) **Pareto Analysis:** this is a statistical technique that delivers a bar chart that shows summaries like the frequency of different issues or input factors. It is a very helpful tool for quickly taking large datasets and presenting them in a clear graphical summary that can direct large audience discussion.
6) **Fishbone or cause and effect diagram:** a fishbone diagram looks to systematically layout the effects of the fault and identify the causes. There are parallels to a FTA but the fishbone has a structure that is very conducive for quality assurance applications.

Brainstorming

These are the criteria for a good brainstorming session based on the Lean Six Sigma process [1].

Fundamentals of Brainstorming

There are several fundamental requirements for a successful session:

1) Everyone must be familiar with the problem.
2) The problem must not be too complex or multifaceted. If it is, a smaller subproblem should be considered.
3) A group of between 3 and 10 people is optimal. A smaller number does not give enough interaction between people so that ideas can feed off of one another. A larger group, on the other hand, is too cumbersome. Some people will be left out of the discussion and will become negative or apathetic – a fatal flaw to a brainstorming session.

Preparing for a Session

One of the most important things to do before a brainstorming session is to define the problem. The problem must be clear, not too big, and captured in a definite question, such as, "What service for mobile phones is not available now, but needed?" If the problem is too big, the chairperson should divide it into smaller components, each with its own question.

Select Participants

The chairperson composes the brainstorming panel. Many variations are possible, but the following composition is suggested:

- Several core members of the project who have proven themselves.
- Several guests from outside the project, but with an affinity to the problem.
- One idea collector who records the suggested ideas.

Draft a Background Memo

The background memo is the invitation and informational letter for the participants, containing the session name, problem, time, date, and place. The problem is described in the form of a question, and some example ideas are given. The ideas are solutions to the problem, and used when the session slows down or goes off-track. The memo should be sent to the participants at least two days in advance so that they can think about the problem beforehand.

Create a List of Lead Questions

During the brainstorm session, creativity may decrease. At this moment, the chairperson should stimulate creativity by suggesting a lead question to answer, such as, "Can we combine these ideas?" or, "How about a look from another perspective?" It is advised to prepare a list of such leads before the session begins.

Three Simple Brainstorming Warm-ups

Brainstorming warm-ups are useful for getting people into the right frame of mind for a session. These three common warm-ups help those involved in the brainstorming process to overcome stumbling blocks and maximize creative results.

- **Word games:** excellent brainstorming warm-ups, word games exercise the mind and help get participants into the proper mindset for brainstorming. It really doesn't matter which specific word games are used, as long as they are mentally stimulating and challenging.
- **Practice run:** brainstorming a completely unrelated topic is one of the more popular and productive brainstorming warm-ups. It is done by creating an amusing imaginary problem and then brainstorming ways to overcome it. Practitioners can get a feel for the brainstorming process and exercise the parts of the brain that will be put to work during the actual session.
- **Game of opposites:** to perform this brainstorming warm-up, write down a list of 10–20 common words. Next to each word, write down the first three words that come to mind when thinking of what the opposite of that word should be. If this is a group brainstorming session, have one person read each of the words aloud as all members of the brainstorming team write down the first three words that come to mind.

Setting Session Rules

While participating in a brainstorming session, there are several rules that need to be followed to make it productive:

- **No criticisms or negative judgments are allowed.** These come later, after the session is finished. The basic idea is to obtain new ideas and not to rate them. The introduction of criticisms, judgments, and evaluations will stop the flow of creative ideas by making individuals defensive and self-protective, and thus afraid to introduce truly new and different ideas for fear of ridicule.
- **Arrange for a relaxed atmosphere.** If the environment is noisy, crowded or full of distractions, concentration will be lost. Also, the positions and personalities of the participants are important. An autocratic supervisor could ruin a session if people are afraid of appearing "silly" and thus do not speak up when they have novel ideas.
- **Think quantity, not quality.** The point of brainstorming is to obtain large numbers of different types of ideas. Again, judgments come later when ideas which do not look promising can be filtered out. By concentrating on quantity, the subconscious is encouraged to continue making new connections and generating more ideas.
- **Add to or expand the ideas of others.** This is not an ego-building contest, but a group effort to solve a common problem. A basic premise is that ideas from one person can trigger different ideas (some closely related and some not so closely related) in other people. That is why this technique works better in a group, as opposed to when used in isolation.

Variations on Classic Brainstorming

Newer variations of brainstorming seek to overcome barriers such as production blocking and may well prove superior to the original technique. The following are some of the alternative options.

Nominal Group Technique

This method encourages all participants to have an equal say in the process. It also is used to generate a ranked list of ideas.

Participants are asked to write down their ideas anonymously. Then, the moderator collects the ideas and the group votes on each one. The vote can be as simple as a show of hands in favor of a given idea. This process is called "distillation."

After distillation, the top-ranked ideas may be sent back to the group or to subgroups for further brainstorming. For example, one group may work on the color required in a product, another group may work on the size, and so forth. Each group will come back to the whole group for ranking the listed ideas. Sometimes, ideas that were previously dropped may be brought forward again once the group has reevaluated the ideas.

Group Passing Technique

In this method, each person in a circular group writes down one idea, and then passes the piece of paper to the next person in a clockwise direction, who adds some thoughts. This is repeated until everybody gets their original piece of paper back. By this time, it is likely that the group will have extensively elaborated on each idea.

A popular alternative to this technique is to create an "ideas book" and post a distribution list or routing slip to the front of the book. A description of the problem should be listed on the first page of the book. The first person to receive the book lists his or her ideas and then routes the book to the next person on the distribution list. The second person can log new ideas or add to the ideas of the previous person. This continues until the distribution list is exhausted. A follow-up "read out" meeting is then held to discuss the ideas logged in the book. This technique takes longer, but allows for individual thought whenever the person has time to think deeply about the problem.

Team Idea Mapping

This method of brainstorming works by using association. It may improve collaboration and increase the quantity of ideas, and is designed so that all attendees participate. The process begins with a well-defined topic. Each participant creates an individual brainstorm around the topic. All the ideas are then merged into one large idea map. During this consolidation phase, the participants may discover a common understanding of the issues as they share the meanings behind their ideas. As the sharing takes place, new ideas may arise by association. Those ideas are added to the map as well. Then ideas are generated at both the individual and the group level. Once all the ideas are captured, the group can prioritize and take action.

Summary

Risk analysis is a cornerstone of just about everything we do. What are the chances of being hit by a car?: Scales 1 (lowest) to 10 (highest).

$$\text{Being hit by a car as I ride my bike on Interstate 95?}$$

$$\text{Severity}: 10$$

$$\text{Likelihood}: 7$$

$$\text{Ability to detect and mitigate before it happens}: 8\left(\text{very unlikely}\right)$$

$$\text{Total RPN}: 10 \times 7 \times 8 = 560 \text{ RPN}$$

OK, how about if I ride my bike on a neighborhood street?

Being hit by a car on neighborhood street Risk Analysis

Severity : 6 (cars go slower)

Likelihood : 3 (road has a shoulder for me to ride on)

Ability to detect car before impact : 4 (they are approaching slow enough I can move over or stop as they pass)

Total RPN : $6 \times 3 \times 4 = 72$ RPN

OK, 560 compared to 72 RPN. Now you don't have to do the math to come to this conclusion. We all do that analysis in our heads instantly when asked if we would rather go for a bike ride on the highway or in a neighborhood. The same thing occurs when an engineer is asked if they would rather use a needle or a ball bearing in an application. But there are significant benefits when we put that question within a formal risk analysis process.

The first is that now the entire team can contribute. It's not possible for the ball bearing engineer to account for every factor associated to the risk, nor do they know every bit of history with bearings in that product.

The second is that by making it quantitative it becomes very easy to turn that analysis into a tool that interfaces with project management and resource allocation.

Yes, tools like FMEAs can take a significant amount of program time. But this is when they are just listed as an activity with hours next to it that hit our schedule. What is not accounted for is that many of these analyses of risk were going to occur informally, and not be listed as a resource consumption. They just secretly bleed program time bit by bit, hidden in design reviews and redesigns.

When we formalize the risk analysis process we are acknowledging:

- It is in fact a program activity.
- It does consume resources.
- If done formally it will be far more efficient and effective.

Simply put "FMEAs save time."

References

1. Wheat, B., Mills, C., and Carnell, M. (2001). *Leaning into Six Sigma: The Path to Integration of Lean Enterprise and Six Sigma*. Boulder City, CO: McGraw-Hill.

11

The Reliability Program

Reliability Program Plan

Many different types of reliability program plans (RPPs) are deployed. Some have been just a cut and paste of a standardized plan, while others only a single hand-drawn Gantt chart on a PowerPoint slide. I'm sure you can imagine I disagree with both of those approaches, but if I had to pick one I would prefer the hand-drawn Gantt chart. A cut and paste of a previous plan is the quickest way to torpedo any program. Doing that repetitively can even sink a department.

The way to develop the correct plan is to start with the following question, "What's the intent of an RPP?" An RPP can have multiple purposes. It's most fundamental is to apply resources to tools that will yield the greatest benefit. "Benefit" can be many things, but this is how I would define it. These are the essential elements.

- Implement a design risk analysis.
- Incorporate guidance from historical information.
- Plan how "measurement of product reliability" will interface with the high-level product program.
- Identify the best tools for improving reliability based on maturity of design.

The reason I am being so extreme about not doing a cut and paste is because nothing hurts an initiative more than wasted investment. Investment in tools that do not yield value is just cause for not funding similar initiatives in future programs: a couple of well-documented examples of wasting investment and it's possible to justify shrinking or even closing a department. It's not too hard to suggest dissolving a reliability department and distributing that functionality among other groups. It can be justified as still being in line with the design for reliability (DfR) philosophy, "We all own and do reliability," but it is not. Without a central command for reliability, DfR falls apart.

So why is a standardized plan a torpedo for an entire department? A standardized plan is disconnected from the actual needs of the product development program. We all know what "one size fits all" really means. This product doesn't fit you or anyone else.

Each product development program is unique in its needs. The idea that program decisions could be made without any knowledge of the specific factors is equivalent to a doctor

Reliability Culture: How Leaders Build Organizations that Create Reliable Products, First Edition.
Adam P. Bahret
© 2021 John Wiley & Sons Ltd. Published 2021 by John Wiley & Sons Ltd.

treating a patient without diagnosis. It's statistically impossible she would do anything that actually addresses the patient's condition.

This is the medical approach from the Middle Ages we now laugh at. "You have a headache?" Bloodletting. "Your finger is infected?" Bloodletting. "You have the flu?" Bloodletting. It was a cure all – or more accurately a "cure none." That process was the result of wanting to do something but not having any knowledge of what to do. Good intentions, fair enough. But it's inexcusable when there are proven techniques for good diagnosis and correct methods that are effective. Using bloodletting in modern-day medicine is malpractice.

The correct approach has tools that matter. This is one of the reasons I created the "Anchoring methodology" (covered in Chapter 8). It is not only important to clearly document "why we need it" and "what is delivered," but we need to keep those updated in a changing landscape.

Using the doctor analogy again. Doctors don't make a plan for a patient's treatment and then just stick to it no matter what happens next with the patient. The doctor will change the treatment based on how the patient is responding. This parallel between a doctor helping a patient and reliability is so effective because they are in fact the same process, "Measuring and improving performance."

Now, it is OK if you have been working from a cut-and-paste plan: don't beat yourself up. It's great that your company has become committed to the reliability process. This chapter will assist with you now evolving to the next level of applying the correct tools from the DfR toolset.

Common Reliability Program Plan Pitfalls

Like anything, there is a list of common pitfalls for RPP. In this section I will cover some of the common missteps that I commonly see in plans and reliability initiatives that can be avoided with a bit of foresight.

The Plan Doesn't Account for a Broad Audience

Don't forget that the RPP isn't just for the reliability group. A good RPP is intended for a broad spectrum of readers. It should clearly explain why each tool or method is being applied. There needs to be a detailed description on how the technique works. There has to be a "why" included. There will be many individuals who look to this plan to understand the reliability initiative.

Some will be a part of the execution, and some own initiatives that engage with the reliability program in some fashion. There will also be readers in the future. They may not have been a part of the program, but will be trying to learn about what has, and has not, worked for programs in the past.

Not Including Return on Investment (ROI)

The activities listed in the RPP should describe return on investment (ROI). It gives some insight into how the selected activities align with the higher-level program and the risk assessment. This can be very helpful when the plan is being scrutinized as time and resources pressures drive leadership looking for fat to trim.

A plan that doesn't have an ROI leaves the reader with the task of independently trying to investigate how tools connect with the greater initiative. That is a burden. I have had to do that with plans I wasn't involved with creating. I can recall taking over a program in this phase of product development. Accelerated Life Testing (ALT) was in the RPP schedule. Why were we doing ALT testing? This was planned almost a year ago, did we still need to do it? Could I cut it because we now had higher-priority tasks that should receive that resource?

It took forever for me to find out why it had been included. I went through the Failure Mode Effects Analysis (FMEA) and couldn't find a high-ranking risk that pointed to premature product wear-out as a concern. I looked through design review notes and didn't see a discussion regarding concerns that would lead to an ALT testing being planned. I ended up in the worst possible situation. No evidence to why it was included and pressured to cut it. I cut it and crossed my fingers. We needed the funding to extend the reliability growth (RG) program.

Too Much

With every unnecessary section in an RPP you risk losing readers. With each section that is over explained, you risk losing readers. When Stephan Hawking, the famous astrophysicist, wrote his bestselling *A Brief History of Time* he was told by his publisher, "For every equation you put in the book you will lose half of your readers." The reason he said this is because the book was meant for a broad audience, not astrophysicists.

If a reader takes great interest in a section of the book, they can research the topic independently. On the other hand, if a reader feels confused by a section, they may put down the book for good. They will never reach that section that truly captured their curiosity and made them want to learn more.

I have seen RPPs that have full mathematical reliability allocation models in them. *I* even felt like putting those down. Why would I be reading an RPP and want to know what the reliability target is for the bearing in the joint "lev12-lower"?

When I need information from the allocation model I'll go find the allocation model. That's the only time an allocation model is interesting, even to a reliability engineer.

But, Goldilocks needs to want to finish it as well.

Too Little

Don't make it so trim that there is no narrative. If there is too little, the plan can seem disconnected. If the narrative has gaps, it will feel confusing. A plan that lists the key activities, like a cooking recipe, doesn't provide understanding. We go to recipes because of higher-level strategies, like hosting Thanksgiving. We want to be creating the *Hosting Thanksgiving* manual.

There must be an introduction. The introduction captures the high-level strategy. This is how the RPP fits into the overall product program. Language should be discussed. Remember, we are appealing to a broad audience that may be new to the subject.

Connect how specific activities are grouped.

- Which activities are aimed at measuring reliability?
- How do the tools/activities connect with the grander product program?
- Which activities are tests aimed at improving sub-assembly reliability?

Not Including Concise and Clear Goals

A plan must include not only goals but also the driving factors they stand on. How were they derived?

An executive once asked, "How can we make a 'perfectly reliable' product?" That's a really bad question. He was told that program to create a perfectly reliable product would be a disaster. The result would be embarrassing market failure that had the potential to put a company out of business. This was not the response he was expecting. I chose to elaborate before he just walked away.

The investment of time, dollars, and labor to create a "perfectly reliable" product would force such a compromise on all other aspects of the product and program that market success would surely be impossible. I can only think of two types of products that could benefit from an approach of trying to create perfect reliability. They would be the Mars Rover and a nuclear power plant. The desire for "perfect reliability" in those cases is driven by either an avoidance of massive loss of life or loss of billions of dollars by a single failure mode.

The Mars Rover need for near perfect reliability I already discussed in Chapter 1. Its functionality could be created by a high-school science club. The extensive development program that went on for years was entirely about reliability. There is no loss of life if the Rover failed. The interest in high reliability was a multi-billion-dollar space program being a waste because a $300 camera or radio transmitter didn't work.

It is reasonable to compromise inclusion of cutting-edge tech, time to market, and development cost in the Mars Rover program. Selling a Mars Rover as a consumer product would not work. This is a product with a billion-dollar price tag and fewer features than some high-end toys. The same functionality can be created in a hobby environment by a college engineering team. It just can't be depended on for a real space mission.

The nuclear power plant reliability is obvious. Failure means tens or even hundreds of thousands of lost lives and devastating destruction to the environment. Chernobyl is a significant amount of land that won't be inhabited again even after a multi-billion-dollar clean-up. The clean-up was simply to stop radioactivity from spreading further around the globe. The root cause of the issue has been partially attributed to a control rod design fault.

The answer to the executive's question is simply, "As a manufacturer of a consumer or general industrial product you don't want a perfectly reliable product. What you want is to create a product with the *right* reliability." The right reliability will be defined at the beginning of the program. It is captured in your product specification document, which dictates the correct balance of product development time, program cost, new features, and reliability based on the product's market objective.

OK, so how do you create a reliability goal that is right? The answer to his question is simply, "Create a program plan based on your business needs and then STICK WITH THE PROGRAM GOALS YOU CREATED" (yes, I'm yelling). A lot of smart people from many aspects of your organization worked hard to create a program strategy that considered all of the product goals and was aimed at long-term maximum market share.

Not Utilizing Testing Initiatives

This one is across the board. Too often, significant investment occurs in test programs with results that are not fully utilized. Some of this can be caused by undercutting test initiatives with delays or underfunding.

Common test execution or usage of results mistakes

- Sidestepping reliability test and analysis for extended feature development. This isn't saving schedule time if we include the program's recovery phase post release due to product issues.
- Ignoring reliability test results because they indicated that the product is not going to meet the reliability target. There is a choice: release a poor-performing product on schedule or take action on the test results. "Head in the sand" isn't an option. That's just dodging accountability.
- Pushing reliability testing out to the end of the program. This will ensure it is almost impossible for any test-results-based actions to be implemented. Saving those actions for version 2.0, to be released next year, is planning for a high warranty hit.
- Don't let a single individual take full ownership of a test. A single owner will tie their success to the test's success. With this we are back to the individual battles that we worked hard to eliminate in the program.

Major Elements of a Reliability Program Plan

Purpose

Starting a plan with a section dedicated to purpose ensures the reader understands why the program is being done. This may be obvious to the core team. A marketing manager may not be very clear as to why we need to do reliability. It's easy to diminish "purpose" in a plan. Resist just skimming over it. The purpose section can identify why reliability is specifically important to this project. Different projects utilize reliability for different reasons. Reliability could be a part of the gain marketing share strategy. Maybe it's the center of a "drive warranty down" initiative.

Purpose (Example)

This document details the RPP for the product. Reliability has been identified as a key element to beating our closest competitor in market share. Customer surveys show that "reliability" is the number two reason for selection of our type of product. We are ranked by customers as only having "mid-level" reliability. It is projected that we can gain 10% more market share by simply improving the reliability of our existing product line without new technology.

This plan defines key program strategy, methods, goals, language, and performance tracking methods associated to the design's reliability initiative. Included are specific design development activities based on previously established areas of risk. This plan will be incorporated into the high-level program plan and accounts for established program goals.

Scope

Scope is needed so the reader understands what will be covered. They may be looking for information that is in a subassembly test protocol or a marketing document. Let's not waste their time.

Scope (Example)

This plan applies to the product development process, manufacturing process, and field usage.

- It establishes the estimated product reliability at the completion of the design on an annual basis.
- Provides methods for reliability testing and analysis of performance data to estimate product reliability over its intended life.
- It addresses product reliability due to failures caused during production or in the field that influences availability to the end user.
- The program metrics for reliability are: failure rate, reliability, useful life, and operational stress margins.
- the tools for reliability improvement and measurement: ALT, Highly Accelerated Life Testing (HALT), Overstress testing, FMEA, Life Cycle Testing, and RG.

Acronyms and Definitions

It stinks to read a document and not know what words mean. These are some common ones in RPPs.

Acronyms and Definitions (Example)

- **Reliability:** the probability that a product will perform a required function, under stated conditions, for a stated period of time or for its useful life.
- **Reliability Demonstration Test** (RDT)**:** a process of demonstrating a reliability metric with a specified statistical confidence.
- **Reliability Allocation:** a top-down process of setting reliability goals for assemblies and subassemblies based on the specified reliability requirement of the delivered product.
- **Failure Mode Effects Analysis (FMEA):** an iterative, bottom-up process, which identifies basic faults at the lowest defined level of architecture and determines their effects at the highest level.
- **Reliability Prediction:** a bottom-up process of estimating the reliability of the finished product by understanding the reliability of its components at a defined confidence limit.
- **Highly Accelerated Life Testing (HALT):** a test method where failure modes are discovered by incrementally applying stresses until failures occur. The stresses may or may not be related to operational stresses. The failure modes are root caused with the intent of being used as robustness design input.

- **Reliability Growth (RG):** the systematic improvement in a reliability parameter caused by the successful correction of deficiencies in item design and manufacture.
- **Out-of-Box Failures (**OBFs**):** units that have passed manufacturing quality assurance screening but are not operational when delivered to the customer.
- **Reliability (Warranty):** this is a measure that establishes a reliability number during the warranty period of the device.
- **Reliability (Life):** this is a measure that establishes a reliability number that covers the established life of the device.
- **Reliability (Maintenance):** this is a measure that establishes a reliability number that estimates the failure rate for life of the device.
- **Environmental Stress Screen (**ESS**):** a quality production test process that uses established over specification stresses to remove defective product.
- **Highly Accelerated Stress Screening (**HASS**):** a tool that screens products at stress levels above specification levels in order to quickly uncover process weaknesses, thereby reducing the infant mortalities, translating to higher quality.
- **Unit Under Test** UUT: the product samples used in the test.
- **Operating Limit (**OL**):** the OL is defined as the last operational temperature or vibration stress point prior to failure.
- **Destruct Limit (**DL**):** the DL is the stress level at which the product stops functioning and remains inoperable at normal operating conditions.
- **Fundamental Limit of Technology:** current technology prohibits the product from working at higher stresses.
- **Unit Functionality Test (**UFT**):** a test protocol that validates a unit has full functional capability.
- **Operational Test:** the test unit is in operation during test execution. This may be accomplished with an automated test script or test operator interaction.
- **Non-Operational Test:** the test unit is not powered on during test execution.
- **Preventative Maintenance (**PM**):** a service procedure that is recommended at set intervals during the product's life.

Product Description

Not everyone knows what the product *is* or how it is used.

Product Description (Example)

This 3 mm diameter and 50 mm long piece of wood is used for removing food in-between the teeth of humans. It is common for users to use it as a way to transport small "bite size" pieces of food such as mini hot dogs, olives, and cut fruit. Customers do tend to use the product in many other creative ways. We do not guarantee any reliability other than the "tooth picking" use case.

Design for Reliability (DfR)

If the program is based on the DfR philosophy, it can be good to explain what it is and why we are using it in this program.

Design for Reliability (DfR) (Example)

DfR is a systematic, streamlined, concurrent engineering program in which reliability engineering is weaved into the total development cycle. DfR ensures that customer expectations for reliability are fully met throughout the life of the product with low overall life cycle costs. It relies on an array of reliability engineering tools along with a proper understanding of when and how to use these tools from the concept stage all the way through to product obsolescence.

The product's DfR strategy will be focused on ensuring true customer usage cases have been characterized and used to establish the product's design parameters.

Performance is measured and communicated based on these parameters. Program tools such as Design Failure Mode Effects Analysis (DFMEA) and Allocation Modeling will guide the team to areas of greatest risk to the design objective. Starting with these tools will ensure the limited resources and schedule will be used in an effective manner.

Once key areas for reliability measurement and improvement have been identified, a focused testing program may be developed. The results of this work will provide a clear set of navigation parameters and the confidence that the product can achieve them.

Some critical program milestones are:

- Clearly defined goals and definitions.
- First assessment of highest-risk design features by end of July.
- Demonstrated first goal statistical confidence by December.

Reliability Goals

The reliability program obviously has objectives. These objectives have to translate to specific product reliability goals. A good product reliability goal is not a specific number like 99.99%. A good goal is a group of terms with specific parameters. It's not excessive; it's accurate. Stating a reliability goal of 99.99% means absolutely nothing if it is not accompanied by the required parameters.

- What time period do I get 99.99% for?
- How can I use it and still get that reliability?
- Is this goal valid if I use it outdoors, in the freezing cold? In the desert in July?
- What is considered a failure against this 99.99%? Do you consider it a failure if I can kick it and it restarts, or all it needed was a reboot? Do paint chips count as failures?

The formal definition of reliability is:

> "The **probability** that a product will perform its **intended function** in a **specific environment** for a given **period of time**."

There are four bold terms in that statement. All four are required for a reliability goal to be meaningful.

- **Probability:** reliability makes no promises. Your specific experience can be anything. We are only aiming at a probability that something may or may not occur. This is not deterministic.

- **Intended function:** the use of the product has to be very clearly defined for two reasons. The first is that it will never work correctly if it is not being used as the designers intended. The second is that even a slight deviation of its intended use can cause damage which means that reliability goal is no longer going to be valid.
- **Specific environment:** a product operates under conditions that we can generalize as "stresses." These stresses come from two sources: how it is used and where it is used. Using a plastic fork as intended, to eat food, can create very different material stresses if it is done at room temperature vs at −20 °C. "Environment" covers all the external conditions that should be considered when defining how a product will perform. It may include, dust, moisture, UV radiation, critters, salt spray, cleaning chemicals, fingerprints, water submersion, nearby equipment gas, or fluid emission. Consider as many as possible. This ensures the promised reliability matches the customer's experience.
- **Period of time:** reliability is time-dependent. By the end of time everything will have a 100% failure rate. If there is no time period for the reliability statement, promising anything over 0% reliability is a lie.

An automobile may advertise a reliability of 99.9%. Does that mean that if I own it for 10 years there is 0.1% chance I will experience a failure? What they probably meant (included in the full statement) is that the car will provide a 99.9% reliability in a one-year period. In addition this reliability is only promised in the first five years. It goes down from there. Year six may be 99.0%. Year 10 may be 50%. Year 15 is 2%. And 2% indicates that you can be very sure you will be having at least one issue each year. It's just about guaranteed.

Reliability Goals (Example)

- The product reliability goal for hard failures is 99.86% annually when used in use case 3 in environment A, B, C. Useful life is 10 years.
- The product reliability goal for hard failures is 99.99% annually when used in use case 1, 2 in environment A, B, C. Useful life is 10 years.
- The product reliability goal for soft failures is 99.99999% annually when used in use case 1, 2, 3 in environment A, B, C. Useful life is 10 years.

Use Case, Environment, Uptime

As mentioned, these definitions are critical for the reliability goal to be communicated. Goal statements were made referencing them. We need to list the specifics and how many exist. They are also fundamental to creating testing program protocols and the interpretation of results. We need specific values to include in the test steps or to interpret test results.

Use cases will include all factors that can be speculated regarding use. What stresses are critical to define when it is used? What is the user trying to accomplish with the product? What are the variabilities in how it can be used?

Environment definitions will include anything that can be considered a variable for a surrounding. Think of a perfect sterile dark vacuum chamber. The environmental definitions include anything different from that environment.

- Is their dust/dirt? How big is it?
- Can it get splashed? How about taken 100 m under water? Is it salt water?

- Does it sit in the sun?
- Can people wear the product and then spray bug spray on?
- Can it be stepped on?
- Can it get thrown in the back of a truck when it's transported after purchase?
- Are we covering the warranty claim if it is frozen?

Uptime is often overlooked when defining a product's reliability profile. Uptime is characterizing the different states a product may be in during its life. The laptop I am writing this on spends part of its life powered on but in sleep mode. When the manufacturer stated it would have a 99.9999% reliability for three years, was that three years of actual usage or three years of calendar time?

Uptime (Example)

- **Non-scheduled time:** time in transit, storage at customer site, storage at manufacturers or distributors site.
- **Unscheduled downtime:** time during failure occurrence, repair, and return to service.
- **Scheduled downtime:** planned time where the product is taken out of service for repair, PM, firmware upgrade, or inspection.
- **Standby time:** time when the product is powered on and not in use as defined in the "intended use" definition.
- **Productive time:** time when the product is in use as defined in the "intended use" definition.

Figure 11.1 shows an uptime stack. This graphic shows how they can relate to each other in accordance with the larger grouping definitions.

Recommended Tools by Program Phase

Planned reliability tools should be presented in reference to what type of initiative they serve. Specifics on the individual tools and how they are grouped "measurement," "improve," and "system" will be included.

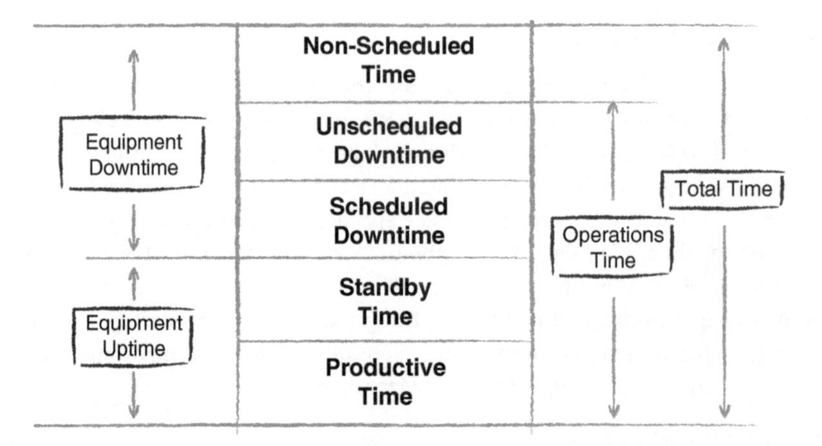

Figure 11.1 Uptime stack.

If the product program has a Gantt chart, it is recommended to include it in this section. The major reliability actions will be included. This provides a high-level picture of when they will occur and what dependencies they share with other program activities. These dependencies are, of course, critical to project managers as task schedules shift.

Design Risk Analysis

We started with "design risk analysis." There are several tools for accomplishing this, but they all share an objective, identify risk, rank them, and take action. The most common are:

- Design Failure Mode Effects Analysis (DFMEA)
- Process Failure Mode Effects Analysis (PFMEA)
- Use Failure Mode Effects Analysis (UFMEA)
- Software Failure Mode Effects Analysis (SFMEA)
- Fault Tree Analysis.

Failure Mode Effects Analysis (FMEA)

There are several types of FMEA's. The most common are Design (DFMEA), Process (PFMEA), Use (UFMEA), and Software (SFMEA). If the FMEA hasn't been completed before the creation of the RPP, the RPP will speculate on the tools most likely to be needed. This can be done with a few quick informal sessions with those closest to the program.

The actual tools and their application will be updated once the FMEA is completed.

If the FMEA has already been completed then a summary of the output should be in the RPP. The FMEA is often the primary tool in the risk analysis. Providing a brief summary of the output helps to further connect how the risk analysis has driven the reliability program strategy. It's not beneficial to dive too deep. If the analysis details are of interest, those readers can pull up the analysis summary report and matrix.

The key elements to include are:

- A brief summary of the highest-risk elements for the product, process, and software.
- The top level – by risk priority number (RPN) ranking – line items from the matrix. It can be helpful to see the logic used in the analysis for the highest-risk items.
- A summary of how the actions items from the top line items connect to activities recommended in the RPP.

Design Failure Mode Effects Analysis, Completed (Example)

The DFMEA highlighted that electronics PC hardware was the greatest reliability risk. The hardware showed a high likelihood of failure due to motion during shipping, and use. Following the PC hardware for high-risk items was the "main pump assembly." With the reallocation of previous main pump assemblies in this new configuration, it is believed that significant vibration transfer to the internals could cause fracture in castings and seals (Figure 11.2).

These identified risks have driven the prioritization of HALT testing for the first prototypes. HALT testing will apply vibration and temperature stress that will identify the specific failure modes of concern. The next available prototypes will be used for ALT testing.

Line item	Index (Option)	Component / Assembly	Failure Mode (FM)	Effect (Performance)	Reliability Severity	Occ	Detect	RPN	Potential Causes
1		PC Hardware (SBC)	Communication failure with internal communications - mechanical disconnect before procedure.	System unavailable, service required.	4	5	5	100	Shipping, shock, jolt dislodge connection
2		PC Hardware (SBC)	Communication failure with internal communications - electrical intermittent/noisy during the procedure	Pressure or heating runs away.	5	5	4	100	Running high power settings, i.e. high flow and max heating creates high-electronic noise environment
3		PC Hardware (HDD)	Corrupted hard drive	System unavailable, service required.	4	5	5	100	Shipping/transport damage (in original shipping container)
4		PC Hardware (HDD)	Reduced performance during the procedure	System unavailable, service required.	4	5	5	100	Intraoperative transport (unpackaged)

Line item	Index (Option)	Component / Assembly	Failure Mode (FM)	Effect (Performance)	Reliability Severity	Occ	Detect	RPN	Potential Causes
9		Main Pump Assy (controller)	Electrical connections to controller	System unavailable, service required.	4	5	5	100	Shipping, shock, jolt dislodge connection
10		Main Pump Assy (controller)	Electrical connections to controller	Pressure or heating runs away.	5	5	4	100	Running high power settings, i.e. high flow and max heating creates high-electronic noise environment
11		Main Pump Assy (controller)	Electrical connections to controller	System unavailable, service required.	4	5	5	100	Shipping, shock, jolt dislodge connection

Figure 11.2 DFMEA table.

Understanding the primary wear-out failure modes will address the next set of identified risks from the DFMEA process.

If a DFMEA has not been completed previous to the RPP creation, an overview of how and when to include one in the program will be outlined.

Design Failure Mode Effects Analysis, Planned (Example)

- The component or function subject to failure.
- The potential modes of failure for the component or function.
- The likely cause of that failure mode.
- The possible effect(s) resulting from the failure.
- The detection controls in place to prevent or mitigate these effects.

An RPN for each potential mode is used to flag critical failures for further evaluation. For high RPN failure modes mitigating design and process actions will be identified and executed during the design development portion of this program.

The DFMEA team will consist of multiple program disciplines. This is a key factor to a successful DFMEA. Required disciplines to attend: mechanical engineering, electrical engineering, quality engineering, compliance engineering, manufacturing engineering, software engineering, and reliability engineering.

The expectation is that the DFMEA can be accomplished in four two-hour sessions with the multidisciplinary team. Analysis and reporting will be completed independently after these sessions by the representative reliability engineer.

Reliability Allocation Model

The model is a tree structure with nodes representing functional subassemblies. The model is not representative of the bill of materials structure. Node relationships are in parent, child, and sibling format (Figure 11.3). Each node represents a testable subsystem or component. Functional nodes contain parts and assemblies that would be joined as a test unit in a RDT.

The model can display multiple model parameters such as reliability, failure rate, mean time between failure (MTBF), use life, and Weibull shape parameter (β). The graphical tree

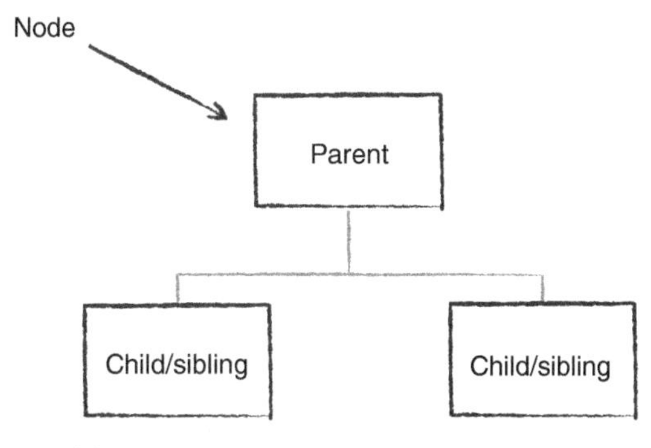

Figure 11.3 Allocation model structure.

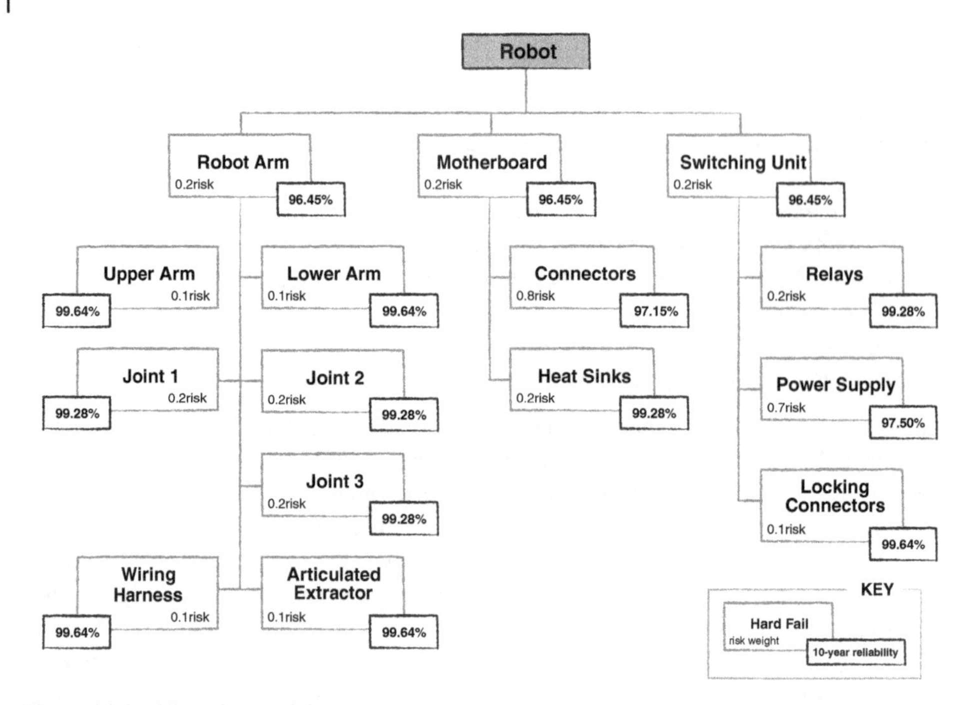

Figure 11.4 Allocation model.

is shown as nodes in a family tree structure. The node values in the graphical tree will update automatically based on the calculator results (Figure 11.4).

The allocation model is constructed by relating subsystems in a family tree structure with risk weighting. Risk weighting represents the likelihood that the component or subassembly will have a lower reliability than its siblings. The weighting is represented as a percentage of the total reliability of the parent.

The sibling's weightings must sum to represent 100% of the parent's reliability. The weighting values are based on the team's knowledge of technical risk and historical performance of similar technology. Weighting values may be adjusted as test results and analysis are completed.

Example: three children (A, B, and C) may have respective weightings of 20, 50, and 30%. In this scenario component B is expected to have the lowest reliability out of the three siblings. Component B is a reliability risk equivalent to both A and C combined for the parent's performance.

Inputs to the model are entered/selected on the model control panel. In this tab product reliability goal will be entered as the "design goal annual reliability hard fail." The units are percentage of failures annually. A separate soft fail goal will be entered in MTBF hours as well. The model may be used to derive component/subassembly goals for either a "hard fail" only mode or a "hard and soft fail" mode.

An environmental stress level may be selected from predetermined values. The three levels are "nominal stress," "high stress," and "extreme stress." the predetermined values have been created based on operating environment evaluations with the team. Uptime percentage will be entered in the model to determine goals for specific customer uptime profiles.

When products have been tested and found to be deficient per the allocation model, there are four strategies to addressing the issue.

- Sibling weighting: if it has been found that a sibling of the deficient design has exceeded their reliability goal then weighting may be adjusted. If the weighting can be adjusted so the deficient design can satisfy a new goal, while not affecting the parent's reliability requirement, these new weightings can remain.
- Parallel redundancy: reliability of an individual component can be increased by including a redundant component that operates in parallel. These two components will deliver the reliability required of the component node.
- Parent or grandparent weight distribution: if parent or grandparent weighting can be adjusted to allocate more reliability to this specific branch, the deficient component may be able to satisfy a new goal. This re-weighting may require re-testing of upstream systems to determine that higher reliability goals for their branches may be met.
- Component/subassembly redesign: the deficient design may be redesigned to increase robustness so it will meet the allocation model goal.

Testing

The planned testing may be locked in or simply proposed for evaluation and selection as the program unfolds. A significant factor in this is what analysis tools and program planning have already occurred when the RPP is created. Remember that the test program is tied to the higher-level initiative by risk analysis and historical data studies. If these haven't occurred then the testing strategy is unlikely to have been finalized. Or, more specifically, the testing strategy should not be finalized. That goes back to the cautions about having a cut-and-paste plan. Keep the test programs tied to risk and the needs of product performance measurement. Maintain integrity with the high-level program.

Organize the testing sections of the program into one of two methods. They can be sectioned either by subassembly testing and system testing, or design improvement testing and measurement testing. Seventy percent of the time they are partitioned by subassembly tests and system tests. This is easier to understand because many people are dividing the program by the timeline. Subassembly tests tend to group together, while system tests occur later in the program.

Grouping by "measurement and improvement" objectives is helpful if the reader is reviewing the plan specifically from the perspective of their role (Figure 11.5). A project manager will be primarily interested in the measurement-based activities, while a designer will be interested in the improve activities.

This is how measurement and design improvement testing differ.

Let's first define what reliability testing is.

> Reliability tests aim to both measure and improve product reliability. Each specific test has a set balance of these two objectives.

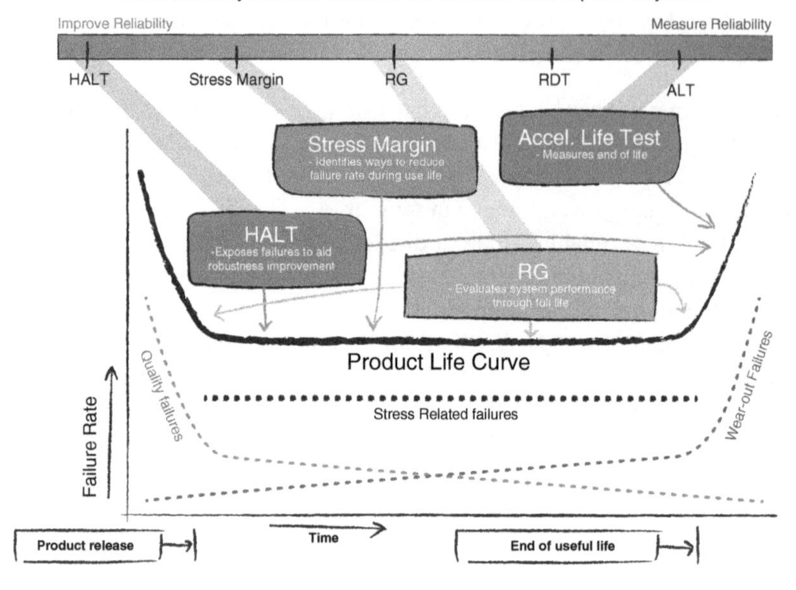

Figure 11.5 Test type by improve and measure ratio.

For example, HALT testing is a 90% improve and a 10% measure balanced test. ALT testing is at the other end of the spectrum with a 20% improve and an 80% measure balance. Figure 11.5 shows this spectrum with a few test types on it in their respective spots. They are also tied to the areas of the product bathtub curve they are most associated with improving and measuring the product.

Highly Accelerated Life Testing (HALT)

HALT testing is one of the most valuable reliability tests, and easily the worst acronym in the reliability engineering discipline. The method and acronym was created by Dr. Gregory Hobbs in the 1980s. He openly regrets calling it HALT testing, and is one of the most daring men I know. He is daring because at a conference I heard him blame his wife, who was in the audience, for the poor acronym. She laughed but I was just like, "How is that drive home going to go?"

This is the problem with the acronym. "Highly Accelerated Life Test," "highly accelerated" is OK because HALT testing applies great stress to show failures as quickly as possible. "Life" makes no sense. HALT testing tells you nothing about the product's life. In fact, its objective is to sacrifice providing any information about life so it can optimize exposing failure modes as input for design improvement. 'Test" doesn't fit because tests are something that we want to pass or use to measure a performance factor. There is no performance factor we are measuring and the only way you could fail a HALT test is if, in fact, no failures were created. When I hear engineers say they "passed" HALT testing, I shake my head because they completely miss the point, and in fact wasted test time.

Another misconception with HALT testing is that it has to be done in a HALT chamber. This is incorrect. HALT testing can be done in any manner in any location the team feels will satisfy the objective. The most extreme I have done is a parking lot with a 1 ft diameter pneumatic hammer. We sold tickets.

I mentioned the objective of HALT testing without really explaining it. The objective of HALT testing is to apply a set of stepped increasing stresses until a failure in the product occurs. The objective is to expose failure modes. The failures will be root caused and used as design input. This enables a rapid design robustness improvement process. This is to some degree a "blind robustness improvement." This term fits because it is difficult to know how much the robustness has been improved until additional tests have been completed.

We often find several design robustness improvement changes we can make that are driven by HALT testing. It sounds like a "loosey goosey" way to do engineering, but it is actually quite brilliant. HALT testing can be done when first prototypes are available, completed in less than a week. The results provide insight into the design that would have taken many months to create with other tests.

HALT (Example)

For the purposes of extended failure mode identification, HALT testing will be executed. The HALT program for the product will be categorized as "Bench Top HALT" and "Chamber HALT." Both categories will share the same objective of "exposing all possible failure modes in the test unit." The Bench Top HALT will be a group of tests that are open to using any stress to be believed to induce failure modes that can be created in the lab. Many of the proposed methods include excessive versions of anticipated operational stresses.

The Chamber HALT category of tests will be based around extreme temperature and vibration stresses. These tests will take place in a HALT chamber at a contracted lab. The lab will use a HALT chamber that accommodates rapid thermal transitions and six-axis broadband vibration.

Testing will use a combination of environments including temperature cycling, random vibration, supply voltage variation, and power cycling.

Electronics printed circuit board (PCB) based assemblies will be subjected to the sequence of low-temperature step stress, high-temperature step stress, rapid temperature transition, random vibration, and combined environments. These test steps will begin within specific environmental conditions and continue until upper/lower DLs are met. In cases where it is necessary to avoid exceeding the DLs of the test asset, testing may be stopped at predefined limits. Figure 11.6 shows the method for applying stepped stress tests.

Subassembly Life Cycle Testing

Life Cycle Testing will provide run time and unit failure data that may be used as input to make statistical reliability and useful life statements. Subassemblies will be run in bench-top HALT tests that emulate normal field operating conditions. The data acquired will be used to determine statistical distribution of failure rate and primary wear-out failure mode time to fail.

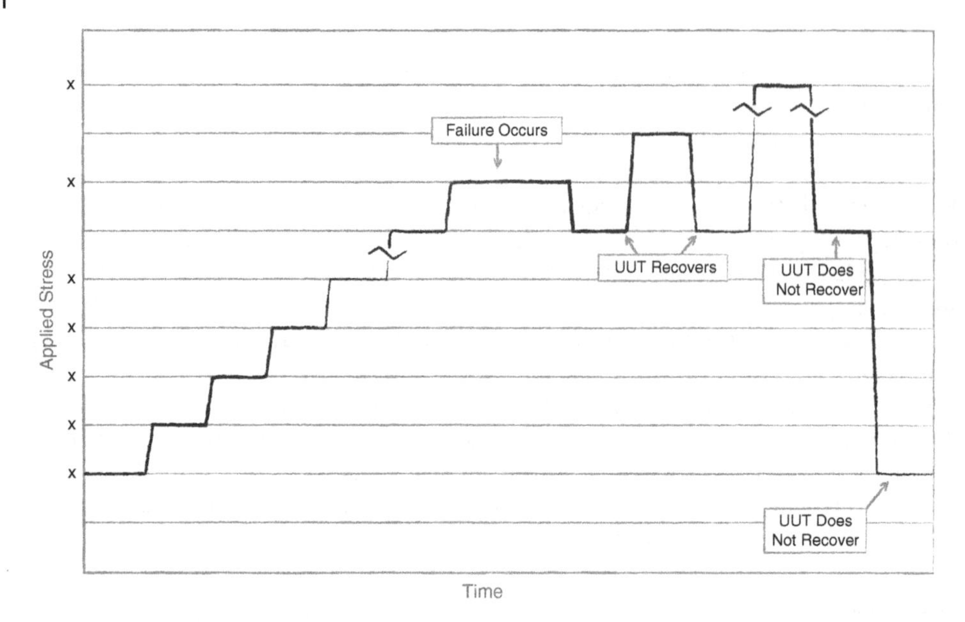

Figure 11.6 HALT testing stepped stress.

Subassembly Stress Margin Testing

Stress margin testing will demonstrate the identified relationship between stresses and functionality that have been identified in the DFMEA and the technical design reviews. Test results will be used to support stress margin statements and provide design input for improved robustness.

Overstress testing will be executed with fresh samples after cycle testing has concluded and corrective actions have been implemented. The mechanical and electrical design team will participate in the test design, setup, and monitoring during execution. Demonstrating margins on identified key stresses is critical to reliability performance. These margins are important to ensure variabilities in manufacturing and usage do not yield failures when in the customer's hands.

Figure 11.7 graphically demonstrates the relationship between operational and design variability. Where these two distributions overlap, failures will occur. Testing focused on increasing the gap between these distributions will increase margin and reduce failure rate. Upper and lower operational limits will be determined in HALT testing.

Accelerated Life Testing (ALT)

Unlike HALT, ALT testing is a perfect acronym. ALT testing aims to create a test to demonstrate product life in an accelerated manner.

ALT testing characterizes product wear-out through a systematic test approach based on the dominating physics of failure of the design.

What is wear-out? Wear-out is when the predictable end of life failure mode will occur. This is how things end for products with perfect reliability. It is a very important parameter to have characterized. Why? Because this is when the reliability you promised to the customer is no longer applicable. Car tires that promise 60 000 miles may advertise a 99.999%

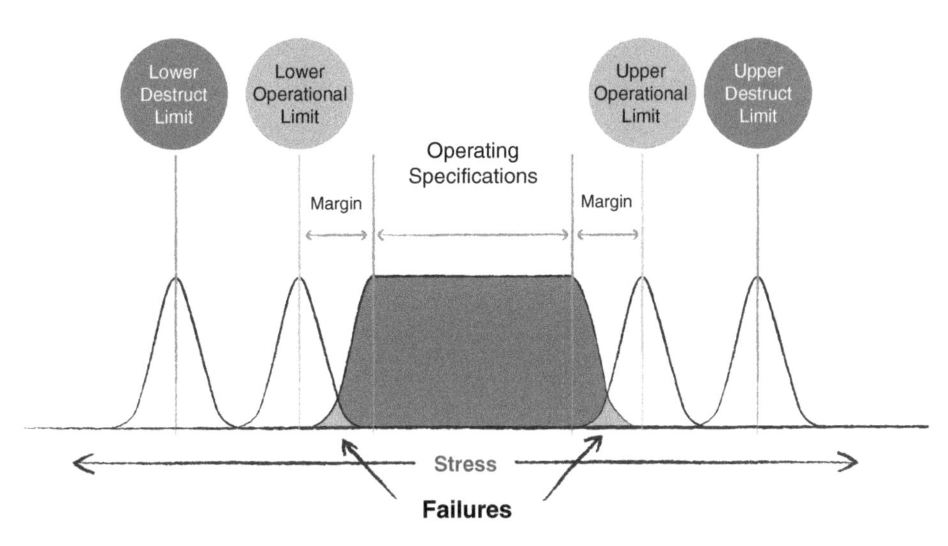

Figure 11.7 Stress margins.

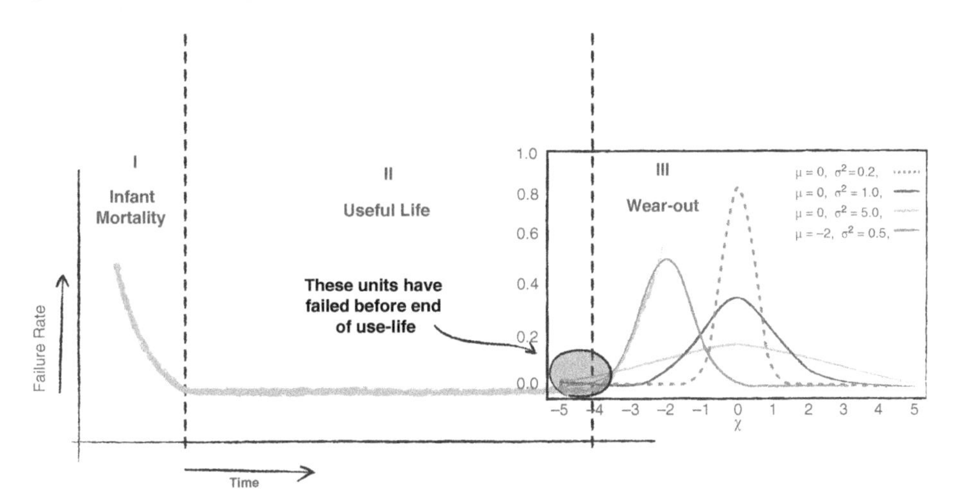

Figure 11.8 Wear-out distribution.

reliability. But not after 60 000 miles. That's why they put that right in the name. If the tire does its job, its life will end with a loss of tire tread – its primary wear-out failure mode. Figure 11.8 wear-out failure modes may be complex. A tire's life can also end simply based on time, not miles. Tires left unused will suffer cracking called "dry rot." You'll find in the small print of your warranty that there is also a limit in time, and not just in miles, for their reliability promise.

ALT testing is accomplished by applying an elevated stress or stresses. The observed failures and the test time required to induce them are translated to use life under nominal operating condition using a model. ALT testing results provide a statistical basis for demonstrating comparative life, compliance with product life requirements, and/or predicting the useful life of the product.

The ALT testing initiative looks to accomplish three things

- Explore the stress to failure mode relationships based on current use case profiles.
- Establish a relationship between elevated stresses and a projected time to wear-out.
- Create a statement on use life for the product based on a specific use case.

One of the most common ALT testing models is the Arrhenius model. This model aims to identify a relationship between the primary wear-out failure mode and elevated temperature. The wear-out failure mode must be based on a material property change through chemical interaction. In many cases this chemical interaction is with oxygen.

An example would be the wear-out of electronics over time. The electronics will ultimately fail due to material property change. This may be a resistance to voltage breakdown or a change in a dielectric properties. By running the electronics in an elevated temperature profile, let's say 30 °C higher than the highest use case, the demonstrated life of five years can be demonstrated in three weeks.

This is clearly a very powerful tool. It lets us see into the future. Like all powerful tools they can create destruction just as quickly when not used correctly. If the assumptions going into the test are incorrect, life statements will be off by orders of magnitude.

System Level Testing

System level testing uses the complete product. The criterion with system level testing is that reliability statements and all interactive stresses are applied as a full use case. What this means is that reliability statements can be directly correlated to what the customer would experience.

Stresses applied in the test should be as complete as possible to ensure that there are no factors that are missing. Sometimes these factors are interactions between individual stresses or conditions which are unknown to the team. But the team can be confident that they have been captured because they have applied the full use case in the test.

My go to system level test is RG. RG will both measure and improve reliability. It is also very adaptable to your specific objectives, a bit of a Swiss Army knife of reliability tests.

Reliability Growth (Example)

System level reliability will be measured and systematically improved by an RG program. The RG team will consist of, at a minimum, members from the disciplines of mechanical, electrical, quality, software, manufacturing, and service. Multiple complete products will be run with runtime and errors tracked.

A selected root cause analysis method will be used to drive corrective actions and design improvements. The Tashiro root cause and documentation method is recommended. The Tashiro method captures all key actions and information for a root cause investigation into a single graphic for review and archiving (Figure 11.9). All internal failures and corresponding information will be stored in the database. RG will be plotted on a graphical chart (Figure 11.10) that highlights key metric parameters. The RG team will select these parameters when the program commences.

<table>
<tr><td colspan="4">ID#: 0xx-</td><td colspan="2" rowspan="2">Owner:</td></tr>
<tr><td colspan="2">Number of Occurrences =</td><td>Impact</td><td>Rev.x.x xx/xx/xx</td></tr>
<tr><td colspan="4">Problem Description (please add any recipe information):</td><td colspan="2" rowspan="2">Picture</td></tr>
<tr><td colspan="4">Analysis:
Why? [Because]
Why? [Because]
Why? [Because]
Why? [Because]
Why? [Because]</td></tr>
<tr><td colspan="6">Root-cause:</td></tr>
<tr><td>Actions</td><td>Owner</td><td>Type</td><td>Status</td><td>Due Date</td><td>Actual</td></tr>
<tr><td>1.</td><td></td><td></td><td></td><td></td><td></td></tr>
<tr><td>2.</td><td></td><td></td><td></td><td></td><td></td></tr>
<tr><td>3.</td><td></td><td></td><td></td><td></td><td></td></tr>
<tr><td>4.</td><td></td><td></td><td></td><td></td><td></td></tr>
</table>

Figure 11.9 Tashiro chart.

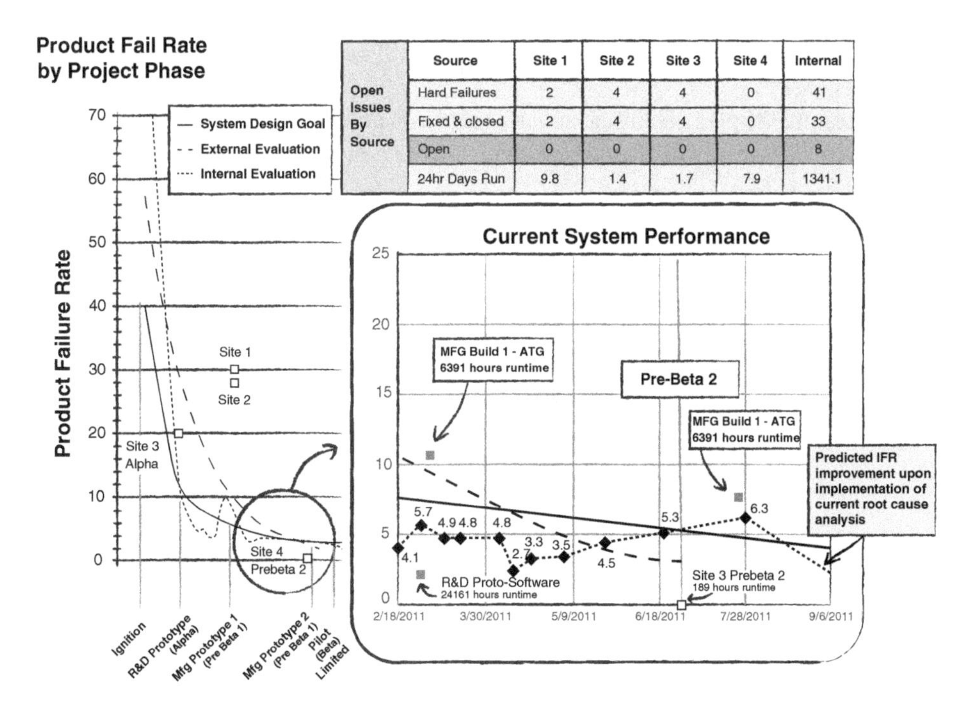

Figure 11.10 Reliability growth plot.

- **Improvement method:** corrective actions based on root cause analysis will be incorporated into the design as the RG process continues. The system performance with the incorporated changes will be validated in designated validation phases.
- **Measurement method:** the reliability metrics will be calculated and reported twice a week. Analysis of data will be directed to program management at requested intervals to aid in program steering/decisions. If RG analysis during development indicates that the product is not on track to meet the reliability goal at launch, project management will be aware of this in advance. This allows project leaders to make informed decisions regarding resources and schedules to achieve program goals.

Summary

It is clear how the RPP provides all the key elements and instructions for conducting the program. A diverse audience can use it as a resource. Each able to extract the pertinent information to their role in the program. A good plan should be able to be read at several levels.

- An executive can read it and take away a clear high-level summary of the strategy and major steps.
- An engineering manager can read it and see what tools are likely to interface with their development programs.
- An engineer can review it and have an idea what will be required for their specific part of the design.
- When creating the plan, include as much of the team as possible. This will ensure it is an informative document and then have ownership in its success.

12

Sustained Culture

Lasting Change

How do immediate changes become lasting changes?

We have all experienced the difficulty of change, personally and professionally. Change is not in our nature: our nature is instinct and habit. For this reason, sustaining change is its own separate initiative.

Stories of failed personal diet and exercise changes are universal.

To be clear, I think taking on lifestyle changes takes courage and hard work. But the image of a guy with the "Vegan for Life" tattoo and a hotdog in his hand (look it up) is the perfect symbol for not treating sustained change as a separate task with its own practices and goals.

Many industry studies indicate that a true change, one that will sustain, is a five-year process. That is five years after the change has been fully implemented.

The assumption is that because a change occurred all forces that direct how we behave have caught up. That's not what happens. The origin of those directing forces are beliefs. When we are acting on beliefs those actions are now our habits. If sustaining the changes in culture we have created is a separate initiative, we need to have a way to track progress.

The Seven-stage Process

One model for this that has been used in different variations over the years shows a seven-stage process. These are the stages:

1) Identifying observational data.
2) Selecting key data from the observation.
3) Adding meaning to the information we have abstracted.
4) Creating assumptions based on the identified meaning.
5) Creating conclusions based on the assumptions.
6) Forming beliefs.
7) We are acting on our beliefs.

Reliability Culture: How Leaders Build Organizations that Create Reliable Products, First Edition.
Adam P. Bahret
© 2021 John Wiley & Sons Ltd. Published 2021 by John Wiley & Sons Ltd.

A more detailed description of these seven stages to creating sustained change would be:

1) **Observation:** the primary task here would be to take enough time to fully observe all pertaining factors. We will ultimately be working with a truncated dataset, but this truncated dataset is most likely to have the correct elements if it is selected from a comprehensive group. We can say that the best practice for stage 1 is "observe thoroughly."
2) **Selection:** we take the pieces of data that we think are most pertinent to our issue or task and separate them. How do we decide what is pertinent? This is actually our first assumption. "I am assuming that this is an important fact but this other one is not." It is necessary to trim down the dataset. It is not possible to continue to work with such a large dataset. What is important is to keep the criteria at the front of our mind. This ensures we are consistent.
3) **Meaning:** this is where it can start to really veer off course. We are heavily reliant on our personal perspective of the situation. "The value of my work is to make the walkways between my garden beds clear." Clear of what? Clear of weeds? Clear of stones? Clear of dirt? This is the first place to take solid action. Once you have your "meaning," share it with another interested party to see whether it matches their interpretation. You may both be incorrect, which may lead to identifying a third meaning to be considered.
4) **Assumptions:** we have taken our meaning and created an assumption. This assumption can be a cause, a state of being, a result, a set of criteria. We placed our own meaning on facts and then placed our own assumptions on that meaning. Assumptions should be shared and cross-checked so a course correction can be made again if needed. Pass them through an eternal checkpoint that is trusted.
5) **Conclusions:** in this step we have now extracted the essence of our assumptions down to a finite set of "facts." There is not a lot we can do with our conclusions once they have been created. The difference between constructive conclusions and conclusions that are not constructive is decided by divining how well the "meanings" and "assumptions" align with what is real.
6) **Beliefs:** this is the final revision of our conclusions that are validated. Put simply, conclusions you are no longer willing to openly discuss and change.
7) **Act:** it is done – automatically and every day.

Summary

The product development process and technology have evolved ever since the first rock became a crushing tool. The speed at which it has evolved has been exponential, or more like a flat curve for two million years and then a vertical line over the past 100.

This is due to the strong correlation between product life cycle and technology growth. How long were spears used before bows and arrows took some of their market share? How long did flip phones dominate the market before smartphones moved in on their customer base? We live in a time where technology is evolving at the fastest pace it ever has, and today's rate of change will look like nothing compared to tomorrow's.

Reliability in design has always been important. No one wanted a bow and arrow that didn't work when dinner was ready to sprint at the slightest noise. But during that time the bow and arrow would be improved in minuscule steps over thousands of years. Reliability wasn't even a process. It was just a natural occurrence. As an invention was used and touched by different hands it slowly improved. For many historical designs it would be impossible to pinpoint when an advancement in a technology actually occurred, like trying to measure a tree as it grows.

As reliability engineering has formalized over just the last century its integration into product development has become more intertwined, and formal. It started with just statistics, then grew into purpose-driven tests, then spread to design practices, and now it is how programs are structured and executed. The "best practices" and formalized methods have to keep pace. That is what reliability culture is about and why the contents of this book are so important. But this book only hits upon 1% of what is happening to advance how reliability practices create great products.

Constantly look outside your projects, your products, technology, and your industry. How we are doing reliability in product development is changing faster than any book or education series can keep up with. It's up to you to see what's new today in reliability and share what you are doing with the world to contribute to the greater good.

Index

Reliability Culture: How Leaders Build Organizations that Create Reliable Products, First Edition.
Adam P. Bahret.
© 2021 John Wiley & Sons Ltd. Published 2021 by John Wiley & Sons Ltd.